ENERGY'S HISTORY

ENERGY'S HISTORY
Toward a Global Canon

Edited by
Daniela Russ
and
Thomas Turnbull

Stanford University Press
Stanford, California

Stanford University Press
Stanford, California

© 2025 by the Board of Trustees of the Leland Stanford Junior University. All rights reserved.

No part of this book may be reproduced or transmitted in any form or by any means, electronic or mechanical, including photocopying and recording, or in any information storage or retrieval system, without the prior written permission of Stanford University Press.

Library of Congress Cataloging-in-Publication Data
Names: Russ, Daniela, 1987– editor. | Turnbull, Thomas, 1986– editor.
Title: Energy's history : toward a global canon / edited by Daniela Russ and Thomas Turnbull.
Description: Stanford, California : Stanford University Press, 2025. | Includes bibliographical references and index.
Identifiers: LCCN 2024048931 (print) | LCCN 2024048932 (ebook) | ISBN 9781503640863 (cloth) | ISBN 9781503641501 (paperback) | ISBN 9781503641518 (ebook)
Subjects: LCSH: Power resources—History—20th century. | Energy development—History—20th century.
Classification: LCC HD9502.A2 E55437 2025 (print) | LCC HD9502.A2 (ebook) | DDC 333.79—dc23/eng/20241227
LC record available at https://lccn.loc.gov/2024048931
LC ebook record available at https://lccn.loc.gov/2024048932

Cover design: Jason Anscomb
Cover art: Alamy. Vintage exposition poster promoting electricity "Exposition Internationale d'Électricité, Marseille April–October 1908," lithograph in colors, 1908, printed by Moullot Fils Aine, Marseille, artist David Dellepiane (1866–1932)

Contents

List of Contributors   vii

Introduction
Toward a Global Canon   1
DANIELA RUSS AND THOMAS TURNBULL

1 "The Largest and Most Important Renewable   21
   Energy Project in the World"
   Maurilio Biagi Filho and the Brazilian Sugar Ethanol Industry
   JENNIFER EAGLIN

2 "Coal Will Be the Primary Fuel of the Future"   35
   Yoshimura Manji on the "Fuel Question"
   VICTOR SEOW

3 The Fear of Being "Left Behind in the Dust"   55
   The Rise and Potential Fall of Coal in China
   SHELLEN X. WU

4 Frederick Tryon and the Decoupling of Energy   71
   and Economic Growth in the 1920s
   ANTOINE MISSEMER

5 The Colony and the World Energy Revolution   84
   Meghnad Saha's Energetic Developmentalism
   ELIZABETH CHATTERJEE

6 The Red Thread to Socialism   103
   *Gleb M. Krzhizhanovskii's "Energetics and
   Socialist Reconstruction"*
   DANIELA RUSS

7 Juan Pablo Pérez Alfonzo and the Invention 124
of Anticolonial Democratic Oil Conservation
MICHAEL DOBSON AND GIULIANO GARAVINI

8 Privatizing a Colonial Electricity Undertaking 144
F. W. Dove's "What People Think of Our Electric Light"
DAMILOLA ADEBAYO

9 Gender, Food, and Vernacular Energy in 157
Moussa Travélé's "Three Rapid People"
LAURA ANN TWAGIRA

10 Uncertain Energy Epistemologies 169
William James and the Case of Mental and Moral Energy
REBECCA WRIGHT

11 Laura Nader's Third-Wave Energy Anthropology 183
THOMAS TURNBULL

12 The Master Resource 202
Energy, Inter-Planetary Capitalism, and
Neoliberal Cornucopianism
TROY VETTESE

Conclusion
Pluralistic Energy History in a Contested Epoch 223
DANIELA RUSS AND THOMAS TURNBULL

Acknowledgments 229

Notes 231

Index 273

Contributors

DAMILOLA ADEBAYO, historian of science and technology with a focus on anglophone West Africa, York University.

ELIZABETH CHATTERJEE, historian of energy and the environment with a focus on India, University of Chicago.

MICHAEL DOBSON, PhD, diplomatic advisor.

JENNIFER EAGLIN, historian of energy and the environment with a focus on Brazil, Ohio State University.

GIULIANO GARAVINI, historian of the Global South, decolonization and energy, Roma Tre University.

ANTOINE MISSEMER, economist and historian of thought, French National Centre for Scientific Research, International Research Centre on Environment and Development (CIRED), Paris.

DANIELA RUSS, historical sociologist, University of Leipzig's Global and European Studies Institute.

VICTOR SEOW, historian of technology, science, and industry with a focus on China and Japan, Harvard University.

THOMAS TURNBULL, geographer and energy historian, Max Planck Institute for the History of Science, Berlin.

LAURA ANN TWAGIRA, historian of technology and gender with a focus on West Africa, Wesleyan University.

TROY VETTESE, environmental historian, University of California, Berkeley.

REBECCA WRIGHT, social and cultural historian of energy, Northumbria University, Newcastle-upon-Tyne.

SHELLEN WU, transnational history with a focus on China, Lehigh University.

INTRODUCTION
Toward a Global Canon

DANIELA RUSS *and* **THOMAS TURNBULL**

HISTORY IS ALWAYS WRITTEN from a specific present. Today, when studying the history of energy use, particularly its role in industrialisation and the mid-twentieth-century acceleration of resource use, such irruptions are inescapably overshadowed by the impacts of climate change. The consequential accumulation of innumerable energy historical events in the atmosphere has radically transformed the meaning of these periods and raises related questions of global politics and justice. For nearly three decades, countries have annually met to assess and negotiate global climate politics at a so-called Conference of Parties (COP), a meeting of the signatories of the 1997 Kyoto Protocol. The original protocol set out a single coordinated emission-reduction plan for all countries—a concerted global energy transition—but this approach failed. Since the Paris Agreement in 2015, each country pledges emission-reduction targets as it sees fit; the meetings serve to review and compare them with the agreed-upon 1.5-degree Celsius global warming limit.[1] The world's uneven, contradictory energetic predicament is among many reasons for the breakdown of this common plan, and this is at the heart of this volume.

The rewards and costs of past fossil fuel use have been unequally distributed in time and space. Within and between countries, those responsible for producing the most carbon dioxide emissions are not the ones who suffer the most from the consequences of global warming. As such, energy historical legacies are an increasingly pressing present-day concern. To take an example, the Sixth Assessment Report of the COP's Intergovernmental Panel on Climate Change (IPCC), out in 2022, referred to colonialism for the first time, noting its role as a historic driver of climate change and acknowledging the specific vulnerabilities it has created for the formally colonized.[2] Moreover, even the dynamics of Earth's atmospheric circulations convey much of the climate's increased volatility toward underdeveloped equatorial regions.[3] Defiantly, Narendra Modi's Indian delegation at the COP persistently affirmed that it would not allow the climatic constraints created by the past emissions of Western and former colonial powers to impede the development of his and other nations.[4] Western and former colonial powers are currently attempting to limit international finance for fossil-fueled development in certain places, such as India, while also outsourcing the material burden of energy transition to formerly colonized places. Accordingly, the term *green colonialism* is now leveled at such strategies.[5]

At the same time, the West's historical responsibility, in part created by its fossil-fueled preeminence, does not, in our contemporary multipolar age, translate into the political power to enforce global emissions reductions. Brazil, Russia, India, China, and South Africa (the BRICS countries) now account for nearly 40 percent of global carbon dioxide emissions, and China has by now become the largest greenhouse gas emitter, on a per capita level comparable to Germany.[6] As we live through and are contributing to a global energy transition, we can see it is a messy one shaped by diverse actors: states, multinational corporations, cartels, state-owned companies, asset managers, banks, and citizens. If we want to understand the specific circumstances under which today's and the future's energy history will be made, we must therefore look beyond Western testimonies.

We have entered an age of energy historical pluralism in which concerned historians must pay closer attention to the global and local forces that shaped a diverse set of energy economies. Our disciplinary subfield has moved on from a focus on Western private multinationals to analyze state-owned companies, whose histories often reach back to decolonization, state-building, and the struggle for sovereignty over resources.[7] Historians have begun to look at the repercussions and meaning of the 1970s oil crises in the East and the South.[8] To further the development of a more representative understanding of the present situation and inform more environmentally sound and historically cognizant futures, our aim is to showcase the diversity of perspectives that have emerged as a result of the pluralization of energy history.

Energy's History hopes to both present and then further this emerging wave of pluralistic scholarship. The book is aimed at students and practitioners of energy history who wish to broaden their perspective or find resources and forms of argumentation that further their research. On one level, we hope the pluralization of perspectives this volume points toward gives all concerned with energy, from scholars to policymakers, pause for thought. Too often energy analysis is based on quantitative or systematized abstraction rather than specification. As a result, simple one-size-fits-all energy policies are proffered as solutions to countervailing problems of energy poverty, development, overabundance, scarcity, and anthropogenic climate change. Where this is accepted, dominant nations impose new forms of what has been termed "energy colonialism" on groups of people and in places for whom the iniquities of the first wave of socio-technological domination are still readily apparent.[9]

On another level, we also hope to further encourage a broadening of the scope of energy historical analyses and their evidential basis. We conceive of this book as primarily an exercise in creating a more comprehensive historical understanding and in presenting a more cosmopolitan account of the various concepts, motivations, resistances, and events that have created our planet of energy users, with all the benefits and costs that this has involved.

Akin to other fields of history, in working toward a more cosmopolitan and global discipline, each chapter presents a primary source and commentary essay that is intended to complexify and broaden the scope of energy history from underrepresented geographic and/or cultural perspectives or from perspectives that challenge most widely held conceptions of energy, even in those places that have been otherwise historically well documented.[10]

ENERGY HISTORY OR ENERGY'S HISTORY

What then do we mean by *energy history*? Broadly, *energy* is a term derived from mid-nineteenth discoveries made in Western European physics and engineering, but one that also borrowed from and leached into other systems of thought, from theories of political economy to moral concepts of work and waste, and the precepts of religion.[11] In its simplest sense, the term denotes the ability of a given system to achieve work. Work, in this sense, is understood as the attainment of change. More loosely, the term *energy* refers to the resources from which such work is derived. The collective noun *energy* denotes anything from hay to human and animal muscle, split atoms, and fossilized carbon. While history, at its simplest, is the study of people, ideas, and things and their ability to affect change over time. So the formulation *energy history*, akin to other specializations like economic or demographic history, implies that the first term is an agent of changes deemed historical. The energy historian thereby chooses to focus upon the history-making capacities of energy or studies events that were shaped, dominated, or determined by energy or even its absence.[12]

Put simply, this emerging discipline begins from the precept that energy deserves to be considered as an aspect of historical explanation. Early-nineteenth-century political economists and engineers debated whether machines made history.[13] In the late nineteenth century, they began to focus more specifically on machines that made nature work and the resources used to fuel them. Arguably, the earliest scholars that we might consider "energy historians" sought to explain industrialism, a notable discontinuity in economic history, as a result of a shift in the composition of the energetic inputs of economic growth. It had been Britain's prodigious use of coal, John Nef argued in his 1932 book, that fundamentally enabled "the ultimate triumph of industrial capitalism"; it allowed the production of goods to exceed the limits of predominantly agrarian forms of industry.[14] Nef stands out for scrutinizing the relationship between energy and economic growth in an empirical manner where others took it for granted. This volume attests to this developmental dynamic in other languages and cultures, some of which we will encounter here.[15]

However, *energy history* should mean more than a history of energy use. Historical theorists have tried to derive the movement of history itself from the two laws of thermodynamics. If history is indeed mediated by energetic laws, progress in energy use necessarily had an entropic exhaust.[16] In 1919 North American historian Henry Adams described the laws of energy as the fundamental engine of historical change. Successful energy conversions powered the advancements of human history, but such processes generated entropy. For Adams, this meant the "ash-heap" of history "was constantly increasing in size."[17] The advancement of historical time was an ambivalent process of increasing efficiency of energy use at the cost of increasing disorder. In the nineteenth-century European scientists had somewhat simultaneously discovered a physical force named energy, but it unavoidably underwent "entropy" over time. In being converted into power, coal becomes ash. Ash cannot become coal. Time's arrow cannot be reversed. So, while ambivalent, energy and history, as Adams claimed, do move *somewhat* in step.[18] However, to too simply impose the laws of energy on historical events risks imposing a "scientifically inflected" form of historical analysis that abstracts and veers into ahistoricism rather than remaining true to our discipline's commitment to historicize and specify.[19]

The twentieth century saw oscillations between visions of energetic progress and decay. The equation of energy and historical progress reappeared in the postwar energy-centered cultural anthropology of Leslie White and in Nikolai Kardashev's categories of planetary civilizations. For White, writing in the 1940s, increased quotients of or efficiencies in energy use were the engine of civilizational progress. While more ambitious still, in 1964 Kardashev argued the most advanced civilization would control and channel the energy of an entire galaxy (a scale that was used in both US and Soviet attempts to discern signals of extraterrestrial life).[20] However, such energy determinists were soon drawn to more earthly matters. With stalling economic growth and increasing environmental awareness, in the 1970s new kinds of energy history appeared.[21] Such perspectives found a new and urgent relevance with the growing if gradual acceptance of humankind's role in the alteration of Earth's climate in subsequent decades.

The reassessment of what it meant to tap into irreplaceable stores of fossil fuels has become more pronounced in recent decades. From the 1990s onward, a newer generation of historians repurposed this evidence to argue that human history, particularly industrialism, had been a fateful preamble to a coming age of a radically new and possibly deleterious age of climate change.[22] Energy history appears then no longer as a story of intentional progress nor natural decay, but as one of human hubris and shortsightedness. Moreover, decades later historians took heed of Earth Scientists and began to argue that both human and natural history has, in a sense, become a field of energy history.[23] The contested notion of the Anthropocene, where humans have become the primary determinant of a transformed planet, hinges upon the amplified agency afforded by fossil fuel use. One estimate is that humankind, in all our heterogeneity, have collectively used 22 zettajoules of "human-produced energy" since 1950. This is more than has been used since the last Ice Age. Moreover, more than 80 percent of this planet-altering agency has come from fossil fuel use.[24] Natural history, science, alone cannot make sense of such an anomalous event. Thermodynamics describes the direction, but it cannot account for the patterns and accelerations we have seen over the last two hundred years.

To understand today's energetic predicament, we must look to human history as much as planetary physics. As William Jordy, a critical biographer of Adams, argued, the historian should rather be "concerned with the path to entropy rather than the destination of entropy itself."[25] These paths are many and varied, as this book shows. In the following paragraphs, we briefly summarize the state of the art in energy history. From the perspectives of the human and social sciences, the uptake of fossil fuels has been typically explained as a result of demographic factors, family structure, economic growth, technological change, imperial expansion, or national or capitalist competition.[26] However, when such conditions are seen as general structures, we risk overabstraction. We can miss what it meant to live through these changes, how people feared, fought, welcomed, and cherished energy—in short, how they made sense of it.

Accordingly, this volume starts from the premise that energy itself must be seen as *more* than a material or technical condition. Energy is an idea that conveys a certain understanding of nature, work, and historical change. As such, the concept emerged amid industrialization and circulated alongside coal and engines. What Reinhart Koselleck called the history of events—in this case structural changes in energy use—did not coincide with the conceptual history of energy.[27] In many ways, the idea of energy precedes energetic historical change: For instance, a high-energy society, abundant and free from toil, was long thought and dreamed about before it was realized in practice. This is the sense in which we would like to think about *Energy's History*: as a relationship between the history of energetic thought *and* the history of energy. So, if the purpose of the volume had to be narrowed down to one argument, it would be that global energy history cannot be understood without reflecting on global intellectual histories of energy.

Rather than adopting energy as a form of abstract historical explanation or zooming out to an undifferentiated species-wide perspective, this book therefore zooms in. *Energy's History* focuses on specific concepts, people, ideas, worldviews, arguments, events, and statements that recorded, critiqued, and informed the distinct energy histories that shape our current and manifold energy predicaments. Each chapter examines a different region, perspective, or intellectual position. In presenting these chapters together, our hope is that much-needed alternative energy futures can be pursued in a manner that accounts for the specific circumstances and motivations that shaped our present reality. We point the reader to the "toward" in the book's title. This qualifier is necessary because we recognize that the creation of a more cosmopolitan and more comprehensive energy history is ongoing.

Moreover, the qualification "toward" also acknowledges that our collective book could not hope to cover all perspectives, places, nor time periods. Here, the focus is on the twentieth century, the period most marked by the global spread of fossil fuel use. Moreover, although we sought cosmopolitanism, three of the book's twelve chapters focus on North America. Among other things, this overrepresentation reflects the contingencies of such a collective endeavor, but we believe this focus can be defended because twentieth-century North America was the greatest energy-consuming nation in this period. That said, this volume should be taken as a start, an incomplete and in-progress task—an invitation to further widen and pluralize our nascent field.

THE STATE OF THE ART

Energy history grew out of an attempt to understand industrial society and its material conditions. This is a question of origins: What is industrial society and when did it form? And one of globalization: How did industrialization spread across the Earth? The distinction between industrial and pre-industrial society is often cast in terms of a transition from an organic crop and wood-fueled society to one that is predominantly fossil-fueled, a distinction popularized by the historians Anthony Wrigley and Rolf Peter Sieferle.[28] However, the cause of this transition to coal in Great Britain is more contested. While Wrigley and Sieferle argue that there were energetic reasons (population pressure combined with a scarcity of wood fuel), others emphasize institutional, economic, or scientific factors, such as the development of patents, the price of labor, and the natural sciences.[29] Historical materialists like Andreas Malm argued that coal was not *energetically* advantageous as compared to water power, for instance, but it improved the control and exploitation of labor.[30] Depending on how these questions are answered, the object, periodization, and political consequences of energy history can greatly vary.[31]

The conventional energy historical story is that the Earth-changing power of fossil fuels was first unleashed in the water-logged tin mines of Cornwall, Southern England. Around 1702, coal was burnt in engines called the "Miner's friend"; with an almost absurd inefficiency these steam-driven pumps raised groundwater from mine shafts.[32] Though inefficient, a feedback loop between Britain's abundant coal, a controlled means of combustion, and the accumulation of capital from producing and consuming energy had emerged.[33] The first patented rotary steam engine, which increased the efficiency and regularity of their power, was presented by James Watt in 1784. Fifty years later, around five hundred such engines were in operation.[34] In using coal, subterranean stocks of fossilized carbon hundreds of millions of years old, rather than wood or charcoal, engines freed industrial growth from the areal limits of solar energy, a quotient limited by the land surface on which sunlight fell and was transformed into plant matter.[35]

Britain's coal-using practices also spread to continental Europe. There, the fuel's combustion first mapped on to the geology of coal seams, residues of once vast forests.[36] However, soon canals, railways, and coal-powered ships used this fossilized carbon to propel coal over land and seas using the mineral's own inherent energy.[37] In 1830, the British East India Company converted its ships from sail to steam power and so powered their contraflow access to the rivers of Asia, Africa, and the Middle East. In the 1840s, the British Navy's steam-powered gunboats then traveled up China's inland rivers, contributing to its victory in the first Opium War (1839–1842) and expediting the extraction of goods outward.[38] While on land, steam-powered railways allowed further incursions inward by fossil-fueled nations.[39] In these ways, coal power transformed time and space on both sea and land, and the expropriation of goods ranging from wood, coal, palm oil, cotton, sugar, rum, became industrialized processes.[40]

That said, archival evidence has long challenged this origin story. Early economic histories of the Industrial Revolution affirmed England as the birthplace of coal-powered industrialism, this was in part because historian John Nef had scoured Western European archives for every trace of this fossil-fueled nation's rise to power.[41] However in the 1960s, one of Nef's students, sinologist Robert Hartwell, uncovered archival evidence that the use of coal and iron on an industrial scale had occurred far earlier, around 1100 BCE, in northeastern China.[42] Comparing records of Chinese coal production to those of Britain, Hartwell showed that far earlier, Kaifeng, the capital city of the Song Dynasty, had established a coal trade considerably larger than that of seventeenth-century London.[43] Nef's British "early industrial revolution" paled in comparison to earlier Chinese industrialism.[44]

Hartwell's evidence overcame the occidental delusion that China had been an inert energetic substrate awaiting the arrival of Western fossil capitalists.[45] With it, the myth of Western coal-powered exceptionalism faltered. Economic historians began the grand task of documenting a longer and far more global history of fossil-fueled industrialism: revisionist highlights included Kenneth Pomeranz's "coal and colonies" explanation for the "great divergence" between the West and the East, and Prasannan Parthasarathi's argument that coal use had a long history in China but that the distance of reserves from the prosperous Yangzi Delta had meant nascent industry failed to catalyze into a system of wider economic feedbacks as it later did in Britain.[46]

Scrutinized more closely, the history of oil poses similar questions. Sujit Sivasundaram has looked to Burma's indigenous oil industry. During the 1824–26 war with Britain, the British were puzzled by the Burmese peoples' use of "Earth oil" as an illuminant, while Britain waged war with coal-driven steamships. Rather than the imposition of Western European concepts and technologies, knowledge about fossil energy had "flowed in both directions," Sivasundaram notes. Why then have historians privileged certain stories over others? What has been lost in most historians anointing North America the birthplace of the oil age, as Daniel Yergin's influential though Anglocentric history of oil did, rather than considering this birth as something that occurred amid the hand-dug artisanal oil wells of pre-colonial Burma?[47]

Amid a wider reckoning with colonial legacies, once-marginalized economic historians are being rightfully reconsidered by energy historians. Long before others, Trinidadian historian Eric Williams explained Britain's outsized industrial productivity not solely as a result of its prodigious coal use but because of the advantages created by combining coal power with the labor of enslaved people. Moreover, Williams pointed out that Matthew Boulton and James Watt's influential steam engine company had been financed by profits from the West Indian sugar trade and that Jamaican plantation owners had been among the first to realize the technology's usefulness as an adjunct to human toil.[48] As Williams put it, the profits of slavery "fertilized the entire productive system" of Britain's Empire.[49] The exploitation of coal and the industry and capital it wrought were founded on the racialized exploitation of labor.

As historian On Barak observed, the "complexities of imperialism are integral to the global fossil economy."[50] Beginning in the port of Aden in 1839, British trading companies built coaling stations across the Middle East to fuel steamships on their way to India and China. The British Navy greatly expanded this worldwide network of naval coal supplies.[51] As such, Britain's industrial revolution was "offshored," creating fossil-fueled colonies in Central America, the Ottoman Empire, India, and British-owned territories in East Asia. The mining, rail, and shipping infrastructure that had been built was used to transmit the productive power of British coal outward and to bring the resources that resulted from this application of power inward, back to this increasingly rich nation.

The completion of the Suez Canal in 1869 connected the Mediterranean to the Red Sea, creating a new and consequential conduit between West and East. Around 20,000 laborers, 3,000 camels, and 10,000 horsepower of steam-powered dredges, paddle-wheels, and excavators—reportedly the greatest concentration of fossil energy assembled on Earth at the time—cleaved a conduit through Egypt. The canal reduced journey times between East and West, and ships carried around 10 million tons of goods a year via it, much of it British coal.[52] Fatefully, in 1892, the *Murex*, a shipping tanker, carried a first cargo of crude oil from Persia to Europe. Fifty years later, around 65 million tons of oil would travel from the global South to North by this canal each year, creating a new and consequential dependency that increasingly assertive oil-producing states would come to exploit in the latter half of the twentieth century.[53]

A global system of fossil fuel extraction and use had been fashioned. Steamships tracked back and forth, fueled and often ballasted with coal, creating a regularized and global energy geography with Britain at its heart.[54] Diesel engines, invented in the 1890s and fueled by distillate oil, increased the efficiency and reach of ocean shipping.[55] The network of coaling stations were the infrastructure upon which later petroleum-driven ocean trade grew, today 90 percent of the world's goods travel by ship, and crude oil constitutes nearly 30 percent of all maritime cargo.[56] As historian Jürgen Osterhammel argued, it is clear that the fossil-fueled age "not only made possible the production of goods on an unprecedented scale but also greatly boosted the formation of networks, speed, national integration, and imperial control."[57]

In the early twentieth century, as industrialism spread, electromagnetic turbines and internal combustion engines became the technologies of a more far-reaching and larger-scale Second Industrial Revolution.[58] As a result inequality among countries' energy use reached its peak in 1913, when Western Europe and North America, with only 20 percent of the world's total population, consumed 60 percent of society's energy.[59] As such, alongside the crimes of colonialism, historians Christophe Bonneuil and Jean-Baptiste Fressoz argue that the "responsibility for climate change of the two hegemonic powers of the nineteenth (Great Britain) and twentieth (United States) attests to the fundamental link between climate change and projects of world domination."[60] Of course, China, India, Brazil, Nigeria, and South Africa, among other nations, similarly pursuing development while at times also seeking to oppose colonialism or dominate others, have since developed means of energy-powered development and have ramped up their rates of energy consumption. Accordingly, understanding the motivations behind *both* fossil-fueled colonialism and its opposition, fossil-fueled and otherwise, is vital if we are to achieve a more cosmopolitan kind of energy history.

A new school has begun the mammoth task of documenting the planetwide and differential reception of energy and its attendant technologies, iniquities, and modes of exploitation. Moreover, the precepts of energy historical inquiry are being widened by area specialists. In showcasing the complexity and diversity of energy historical sources, *Energy's History* hopes to challenge an overly simplified center-periphery conception of the spread of energy history. The contributions in this book are organized into three distinct thematic sections. The first concerns the relationship between energy and development. The second section addresses the emancipatory potential of energetic development in colonized, anti-colonial, and communist contexts, while the third focuses upon contested intellectual histories of energy.

DEVELOPMENTAL ENERGIES

Each chapter sets out an account of energy-powered development, achieved either by imitating Western coal-based development or by using alternative fuels, or less fuel. In pursuit of a more cosmopolitan energy history, yet another objective of this volume is to draw attention to a growing number of histories of alternative energy.[61] In her contribution, which opens the volume, Jennifer Eaglin outlines arguably the world's largest experiment in alternative energy to date: the Brazilian state's plan, from 1954 onward, to convert the nation's growing fleet of cars to run on sugar-derived ethanol. The Petrobras program potentially prevented around 14 billion liters of gasoline from being burnt.[62] Unfortunately, this "Brazilian Arabia" exploited rural labor, degraded the environment, and created unmanageable quantities of foul-smelling pollution. In chapter 1 here, Eaglin presents a 1983 appeal for the program's continued support from sugar magnate Maurilio Biagi Filho. Eaglin's essay and the book on which it draws present a cautionary tale for those implementing today's low-carbon energy transitions. While such changes may become necessary, it appears that alternative fuels always come with alternative costs.

Capitalist, communist, or other, energy has long been enrolled in developmental nation-building projects. In his book *Carbon Technocracy* Victor Seow took a coal mine in Fushun (today Northeast China) as a specific site from which to explore energy history amid distinct and shifting political regimes: Chinese nationalist government, Japanese Manchurian rule, and then Maoist communism. All three regimes enacted a kind of planning-intensive carbon-exploiting "technocracy," which lends the book its title and challenges any simple relation between fossil fuel use and liberal capitalism.[63] Drawing on this work, in chapter 2 of this volume, Seow presents an essay by Yoshimura Manji, founder of the Japanese Fuel Society. The source is a foundational text on state coal use. Manji set out a developmental "fuel question." Could Japan imitate European and U.S. conservation and scientific management efforts or even surpass them? In presenting the developmental vision of a latecomer to coal use, Seow shows that fossil fuel use was seen as offering a means of resistance to dominant global powers as much as a means of ensuring those same powers' growth.

With similar care, Shellen Wu's *Empires of Coal* (2015) documents China's turn to coal, a process that involved the reappropriation of European geological knowledge in the early 1900s.[64] In chapter 3 here, Wu provides a foundational text by state minister Tian Wenlie, who used his preface in China's inaugural *Bulletin of the Geological Society* (1919) to argue that the late Qing dynasty should fully exploit China's coal seams and not cede this fuel to Europeans. A foreword from Ding Wenjiang, who directed a survey of China's coal reserves, is presented alongside the preface. In both sources we see the roots of the state's justifications for China's current coal-powered economic strength and geopolitical authority. However, Wu argues that in an age of global heating, the sustainability of this arrangement has begun to falter. The state now hopes to become an alternative energy superpower while still retaining the political power that coal afforded.

Yet, despite the pursuit of energy-driven development, the relation between energy and development remains uncertain. In exploring this in chapter 4, historian Antoine Missemer presents the work of forgotten North American energy economist Fredrick Tryon, which showed how decoupling, the belief that energy consumption can be separated from economic growth by market forces or political interventions, was not an idea fashioned amid the post-1980s sustainability agenda but in the Progressive Era United States. Missemer argues that the mechanism of decoupling was an outcome of straitened circumstances and conservation measures implemented by the United States during the First World War. Today, given that the IPCC puts great faith in the potential decoupling of energy from growth, the idea that energy may be less strongly coupled to economic development than previously thought should draw upon such historical precedents in helping to formulate policies for reducing energy use.

EMANCIPATORY ENERGIES

Many nations, On Barak notes, "first encountered steam power while facing the barrel of a gun." However, the Ottoman Empire and China, in 1830 and 1860 respectively, began to develop their own steam-powered engines. Fossil-fueled domination soon became fossil-fueled resistance.[65] It was the West's connection to oil that would prove the most consequential means for implementing fossil-fueled anti-colonialism. The Suez Canal created a channel from the Mediterranean to the Red Sea, an artery through which

Europe became hooked on Middle Eastern oil.[66] When US oil production fell in the postwar period, petroleum-rich states increasingly sought self-determination and profit. Historian Giuliano Garavini has rightly argued that the formation of the Organization of the Petroleum Exporting Countries (OPEC) in 1960 created the "first international organization of the so-called 'Global South.'"[67] The imposition of oil price rises and embargoes that followed were a kind of anti-colonial oil weapon deployed against comparatively rich oil-consuming nations who had once profited from the subjugation of others.

Even before the formation of OPEC, as historian Elizabeth Chatterjee has argued, non-Western nations were never simply passive receptacles for the supposedly innate logic of fossil-fueled growth. Many nations proposed alternative trajectories in the early twentieth century, a process she characterizes as "fossil developmentalism."[68] As such, her work suggests energy history was less a shockwave that spread from Western Europe than a set of behaviors that now rippled back and forth and encouraged counter-hegemonic visions of energy-powered development. In chapter 5, Chatterjee recounts the vision of the Dalit anti-colonial physicist Meghnad Saha, who in 1944 put forward the idea that India's rivers and potential atomic power could achieve a distinct energy-intensive modernity of a kind that had been denied under British colonial rule. Saha's plan subtly reflected Indian perspectives regarding caste, labor, and anti-colonialism. Moreover, in asserting India's right to energy-driven development, in contrast to a Gandhian low-energy vision, Saha presaged later discussions about the right of formerly colonized nations to emit carbon dioxide and other emissions just as their oppressors had.

In a sense, the Soviet Union can be seen as a large-scale experiment in energy-driven anti-hegemony. In chapter 6, historical sociologist Daniela Russ presents the forgotten work of Soviet electrical engineer Gleb Krzhizhanovskii, who was convinced a prosperous socialist state could be achieved only by placing energy, particularly electrification, at the center of economic planning. As head of the state planning commission, in 1929 Krzhizhanovskii published a vision of the Soviet's energy future in which labor, thermal, and electric technologies would harmoniously ennoble "man" and his environment. Unlike capitalism, socialism could be planned according to energetic considerations and evolve more efficiently. However, Krzhizhanovskii was sidelined by Stalin. The fossil-fuel-intensive Soviet rule that followed provided a prehistory of the region's present environmentally and geopolitically disruptive energy politics and the centralized regime it supports. Dreamed-of energy futures are not always realized, but such visions help coordinate action in the present.

This thematic section also includes a chapter by Michael Dobson and Giuliano Garavini on the highly influential anti-colonial approach to oil conservation formulated by OPEC founder and Venezuelan oil expert Juan Pablo Pérez Alfonzo. Minister for Petroleum between 1959 and 1963, Pérez Alfonzo witnessed how his and other developing nations were becoming the world's predominant oil suppliers and realized that this was an opportunity to redress iniquitous colonial relations. We were able to secure the right to reproduce extracts of his key text *El Pentágono Petrolero* (1967) in chapter 7. This manifesto sets out strategies for regaining self-determination over oil, a suite of policies for its taxation, conservation, and nationalization that was intended to fight the exploitative concessionary agreements oil-rich countries had been coerced into in the early twentieth century. Dobson and Garavini describe how Pérez Alfonzo had recognized the risks that oil posed—the inequality, corruption, and environmental damage—but he also saw oil's potential as a liberatory resource, a pragmatic means for assuring independence. The conflict between development and the damage this entails for others remains central to energy history.

However, not all anti-colonial fossil-energy projects have advocated for state control. In chapter 8, historian Damilola Adebayo takes us to colonial Nigeria. There in 1909, elite West African barrister Frederick William Dove hoped Lagos's citizens could benefit from electrification, but he argued that the private sector rather than the state should lead development. Dove's call for privatized electricity supply, Adebayo argues, constitutes the first documented call for the privatization of electrical infrastructure in Africa. This call provided a prehistory to twenty-first century debates in Nigeria (and elsewhere) about the right way to provide electricity. Adebayo notes that in a sense, Dove presaged later critics of state-run energy industries and their supposed tendency toward inefficiency and waste. Today, continued underinvestment in Nigeria's state-funded electrical infrastructure suggests Dove's call for private funds may have been prescient, in this context at least.

IDEATIONAL ENERGIES

At its core, this volume is concerned with how people have thought about energy as a precondition for action. Energy is an abstract concept, but it is related to the specific ways in which humans experience the limits and affordances of the environment and their bodies. Alongside the well-known Aristotelian *energeia*, Indian-Buddhist traditions suggest that *prana*, a Sanskrit term indicating breath, is akin to energy. In the Chinese-Daoist tradition, *qì* (気) refers to a life-force permeating all beings and granting the universe flow.[69] In the 1900s, physician Tang Zonghai coined the term *qihua* (氣化, qi-transformation), which referred to steam engines, to explain changes human bodies were capable of.[70] Rather than being "against energy," it is more ecumenical to recognize coherence between non-Western instantiations of energy-like thought.[71] Energy and energy-like concepts relate to universal human problems: the hunt for warmth, nutrition, freedom, and survival.

In chapter 9, Laura Ann Twagira provides an example of such kinds of vernacular thermodynamics by recounting a Bambara folk tale "Three Rapid People" from the Soudan region (today Mali). The story concerns *nyama*, a Mandé term for a variety of forces that can be harnessed by humans, which might be thought of as a vernacular energy concept. Recorded by interpreter Moussa Travélé, the story describes how *nyama* affected imagined acts of harvesting and cooking millet and hunting, at speeds that defied belief. In describing a supernatural force that could alleviate human labor, the Bambara tale is part of a seemingly global genre, that of "power fantasy," imagined situations in which energetic limits and entropy no longer apply.[72] However, the story focuses on traditional labor at a time when fossil-fueled French colonialism had begun to alter long-standing rhythms of Soudanese domesticity. Colonialists' own power fantasy was that the diet of their local workforce could be optimized via measurement and control of their calorific intake. In Twagira's energy-historical reading, "Three Rapid People" reflects a long-standing and universal desire for plentiful and labor-saving forms of nutritive energy.

In Rebecca Wright's contribution, chapter 10, we learn that in the West, where energy was influentially defined by natural scientists, aspects of the concept still remained unexplained. In 1909 North American philosopher William James, Wright explains, published a philosophical paper on the notion of "mental energy," an individual's perceived sense of motivation. We are all familiar with this sense of willpower, a kind of "imperfect vitality" that escapes precise definition and measurement. James considered mental energy a precondition of all human action and the means by which conventional stocks of energy were actualized and put to work. For James, the riddle of mental energy put paid to Henry Adams's idea that there was a clear and progressive relation between energy use and history, as not only was mental energy an imprecise variable, but it was one that, without moral guidance, could result in destructive outcomes. This prophecy about the misuse of energy resources would soon to unfold on the mechanized and fossil-fueled battlefields of World War I. As Wright argues, James emphasized the uncertainties that still surrounded the energy concept and challenged its use as a meter of social progress.

Thomas Turnbull's chapter, 11, returns to the long-standing idea of energy determinism, the idea that energy is the ultimate determinant of historical change. He discusses a 1981 lecture from the Berkeley-based anthropologist Laura Nader. The text was an outcome of ethnographic observation of U.S. energy policymakers in the 1970s. Nader had documented how a prejudicial gendering of energy technologies by predominantly male energy analysts acted as a barrier to deploying new and alternative energy technologies. Moreover, Nader's anthropological experience fed into her argument that there was no evidence that increased energy use was correlated with increased quality of life. Instead, she made the case for a value-driven energy policy based on desired lifestyles, decentralization, and alternative technologies. Having done so, Nader resolutely challenged an anthropological legacy of energy-based social evolutionary thought, of the kind put forward by Leslie White and others, in which imagined social hierarchies found credibility through recourse to differences in energy consumption.

In the book's final chapter, 12, Troy Vettese addresses the idea that our energy supply is essentially unlimited, so the extraction and use of energy should not face any constraint. This power fantasy is part of a distinct neoliberal conception of human-energy relations. Vettese presents this idea as a facet of US economist Julian Simon's wider philosophy of "resourceship," which he promoted in that tumultuous decade for energy history, the 1970s. Simon argued that if market forces were allowed to act in an unimpeded manner, the laws of energy place no real limit on human flourishing: the true catalytic potential of human ingenuity and the physical potential of our and even other planets' resources could be fully realized. Today, such cornucopian thinking is taken as gospel by many. As demand increases, new forms of unconventional fossil energy are found, and the continued availability of new energy resources is, for now, not in doubt. However, the ultimate limit now appears to be our planet and the boundaries that are imposed by its various environmental parameters.

A VOLATILE CANON

In an age of global heating, persistent petrochemistry-derived pollution, sea level rise, and mass extinction, energy history has become a largely tragic tale of human hubris, excess, and unanticipated environmental outcomes. We must increasingly recognize not only how human history has been altered by fossil-fueled industrialism but also consider how energy has been a central driver of fundamental changes to human and what was once called natural history. This is not a simple exercise in historical revisionism; a more expansive energy history can offer examples and frames of meaning that can help inform the actions of those looking to transform contemporary energy systems and their attendant political relations. Which is to say, advances in energy historical understanding are of great consequence for energy futures.

Not every author from this increasingly cosmopolitan energy-historical turn is featured in this volume. Readers must look elsewhere for recent accounts of Mexico's adoption of oil or its electrification.[73] Historians have documented the electrification of Late Imperial and Early Communist Russia, Revolutionary China, and Israel-Palestine, and still others have told of the imposition of Anglo-American fossil-fueled capitalism in East Asia, the Soviet-influenced Ghanaian nuclear power industry, the effects of the OPEC oil embargo in Tanzania, and the development of oil infrastructure in Iran.[74] Abigail Harrison Moore and Ruth Sandwell have affirmed women's centrality to energy history.[75] The field is becoming more geographically and socially representative, a development we hope our collective efforts can contribute toward.

The perspectives this volume presents are incomplete and are no doubt a reflection of existing networks, circumstance, our own specializations, and—admittedly—biases. However, together, they are nonetheless intended to give readers a sense of the possible scope of energy history if it is approached from a more pluralistic perspective. Energy historians should be more ambitious in delineating the scope of our nascent discipline and more imaginative in deciding what are appropriate sources for energy historical inquiry. Rather than being against energy, we argue that we should begin to think with it anew.

That said, to those who assign too much explanatory power to energy, we counter that while it is emphatically an aspect of historical explanation and one likely to become increasingly apparent it is not an agent of history in and of itself. Energy does not explain history, but the innumerable events its exploitation has resulted in demand modes of explanation that are rightly inflected by the strictures of energetic laws and the thoughts and actions that have developed around them and as a result of their use. We hope these essays and accompanying sources provide one starting point for a more expansive, cosmopolitan, and globally representative energy historical canon. Unlike those of the past, this canon is imagined as a dynamic repository of major and minor texts, a library without hierarchy, whose cataloguing system is perpetually contested and incessantly reshaped as the ever-advancing present imposes new demands upon our shared historical understanding.

1 "The Largest and Most Important Renewable Energy Project in the World"

Maurilio Biagi Filho and the Brazilian Sugar Ethanol Industry

JENNIFER EAGLIN

SUGAR AND ETHANOL PRODUCER Maurilio Biagi Filho resoundingly asserted Brazilian sugar ethanol's importance in a 1983 opinion piece, opining that "every Brazilian feels proud and responsible knowing that here, in our Country, with technology and manual labor of Brazilians, we were able to successfully develop the largest and most important renewable energy project in the world." In November 1975, Brazilian military-president General Ernesto Geisel had implemented a national ethanol program, Proálcool. This state-led development program focused on expanding Brazil's sugar-based ethanol production in order to economize on foreign currency through the substitution of petroleum imports; reduce regional and individual income disparities; increase internal income through the expansion of domestic jobs; and expand the production of capital goods through "highly nationalized" equipment contracts in the expansion, modernization, and establishment of ethanol distilleries.[1] Implemented in the midst of a global oil crisis, the program transformed the sugar industry into a large-scale, modern alternative fuel producer. By the 1980s, the program supplied ethanol to mix in the national fuel supply and to power domestically developed engines that ran exclusively on ethanol. However, critics lambasted the high financial, social, and environmental costs associated with the program and the industry it supported. In response, the prominent sugar and ethanol producer and outspoken ethanol promoter, Maurilio Biagi Filho, wrote an impassioned defense of the ethanol industry, "O álcool é nosso" (The ethanol is ours), to silence critics and garner public support for the embattled program. Today, Proálcool remains one of the world's largest and most successful projects to have integrated renewable energies into a national energy system.

This essay places the Brazilian ethanol program in the context of wider domestic sugar interests and a national development agenda, with the aim of presenting a widely unknown history of renewable energy that may help to inform contemporary moves toward a non-fossil energy economy in a manner both encouraging and cautionary. Previous historians have presented the origins program as simply a response to the global fuel crisis of the 1970s,[2] but the Brazilian government had been investing in ethanol's development for decades prior to this. Because Brazil had only limited domestic oil reserves, interest in ethanol as an alternative to petroleum had driven both government and private investment in sugar-based ethanol since the 1930s. In the 1970s, private producers like the Biagi family seized the opportunity and tied their agro-industrial dreams to the military government's energy objectives. Together, private producers and government incentives successfully integrated ethanol into Brazil's energy infrastructure through Proálcool, a move that continues to have multifaceted impacts on both Brazil and the world's alternative energy options to this day.

Ethanol, or ethyl alcohol, is distilled from any starchy agricultural products, be they potatoes, grapes, corn, or sugarcane. Fermented sugars from these starchy products can be distilled or refined into drinking alcohol. Low-grade distillations produce liquors while high-grade distillations are potent enough to run engines on. The technology to use ethanol as a fuel supplement or fuel replacement has been around as long as the internal combustion engine. Challenging still ongoing assumptions that fossil fuels are superior to renewable fuel options, early supporters of the use of ethanol rather than petroleum included moguls who later became central to the advancement of fossil-fuel-based fuels and electricity infrastructures such as automobile magnate Henry Ford and electrification booster Thomas Edison.[3] Despite the later ascendance of fossil fuels, as countries without large oil reserves, France, Germany, and Brazil all funded research into alternative ethanol fuels.

State-sponsored Brazilian research in sugar-based ethanol formally began in the 1920s as the Brazilian government, along with the rest of the world, increased the attention given to domestic fuel sources as a result of shortages during World War I.[4] In 1923, São Paulo engineer Eduardo Sabino de Oliveira conducted research on this potential energy source in collaboration with researchers at the São Paulo Polytechnic School and the National Technology Institute in Rio de Janeiro under the Ministry of Agriculture's former Mines and Combustion Experimental Station. Their findings proved that of the two basic ethanol varieties, anhydrous and hydrous. Anhydrous, or waterless, alcohol, as it was referred to in Brazil, mixed best with petroleum fuel at rates of up to 25 percent, which allowed for efficient combustion without having to adjust the engine.[5]

With the onset of the Great Depression, earlier state-sponsored interventions, alongside the collapse of agricultural export markets, including that of sugar, encouraged Brazilian government officials to formally promote ethanol use at a national scale. Beginning in the 1930s, the government began mandating a 5 percent mixture of ethanol in the national fuel supply.[6] This small-scale mixture supported the sugar industry by redirecting excess sugarcane from the export market toward domestic ethanol consumption, helping also to diminish oil imports. While controversial at times, as we shall see, this mandated 5 percent mixture would remain in place for the next forty years until an energy crisis increased the attention given to alternative fuels.

International politics dramatically changed the global oil market in the early 1970s. Following American military support of Israel against Egypt and Syria in the Yom Kippur War in October 1973, Arab countries aligned in the Organization of the Petroleum Exporting Countries (OPEC) to impose an embargo on oil exports and cut back overall production by 25 percent to assert their anti-imperialist position against the United States and other nations.[7] Within months, global petroleum prices quadrupled. As a country that relied heavily on foreign oil (80 percent by 1973) and that still had limited domestic oil reserves, Brazil proved particularly sensitive to this global price shift. Furthermore, a conservative faction within the military had established a Brazilian military dictatorship nearly a decade earlier, which had promised to solve Brazil's rising inflation problem and to support economic growth. After six years of record growth, known as Brazil's "Economic Miracle," the recessionary implications of increased oil prices threatened this promising economic growth.⁸ Under the military dictatorship, officials sought solutions to impending energy and economic crises generated by the oil price hike. Yet, even though the government had long invested in ethanol as part of domestic energy policy, this fuel was initially only a minor option among many other diversification options such as hydroelectricity and nuclear power.⁹

Private producers like Maurilio Biagi Filho were critical to turning the government's attention toward a nationalist large-scale ethanol program. The Biagis had migrated to Brazil to work the coffee fields of São Paulo, like a million other Italians in the late nineteenth and early twentieth centuries. Like so many others, Pedro Biagi worked off his debts and bought coffee lands to amass considerable agricultural wealth through selling this enervating bean. His son, Maurilio Biagi Sr., bought a local sugar mill, the Santa Elisa, with his brothers and a business partner in the 1930s and they began establishing themselves in São Paulo's growing sugar industry. By the 1970s, the Usina Santa Elisa, headed by Biagi Sr. with Biagi Filho, became the center of the family's agro-industrial sugar and industrial-equipment manufacturing empire.¹⁰

In the wake of the oil crisis, Biagi Filho and his father sought to promote a policy, both in open debate and behind closed doors, that would increase the quantity of ethanol in the national fuel mixture.¹¹ When the government instituted the formal program, the Biagis were among the first producers to win project funding. By relying heavily on expanding already existing distilleries owned by producers like the Biagi family, domestic ethanol production grew from a little over half a billion liters of alcohol per year in 1975 to three billion liters of alcohol by 1979, and this growing resource was used to increase the amount of ethanol mixed in the national fuel from 5 to 20 percent.

Sugar producers' role as energy producers dramatically expanded in the late 1970s and 1980s when the government retooled the program to become a full fuel replacement program after a second oil shock and following the Brazilian launch of the ethanol-fueled car in 1979. The beginning of the Iranian Revolution in December 1978 sparked a 14.5 percent increase in OPEC's prices and provoked production cuts, after which posted oil prices nearly doubled by the end of 1979.¹² This political turn in Iran, Brazil's second largest oil supplier, left Brazilians even more vulnerable to the new crisis than the first shock had.¹³ In the face of rising oil prices, officials focused on expanding the ethanol program.

The Brazilian ethanol-fueled automobile was a source of nationalist pride for Brazilian policymakers, military officials, sugar producers, and ethanol consumers alike. Running exclusively on ethanol required numerous adaptations to the standard gasoline-run Otto engine. Brazilian engineer Urbano Ernesto Stumpf and his team of researchers at the Aeronautical Technology Center (Centro de Tecnologia Aeronautica [CTA]) developed the modifications, adapting gas turbines, compression ratio, carburation, ignition distribution, and the cold engine start system. The CTA launched the first fleet of experimental ethanol-fueled cars for government use in 1977. Although foreign companies like Volkswagen had also worked on ethanol cars, it was Brazilian scientists who won the patent on the engine.[14] For some, such as Biagi Filho, winning the royalties for ethanol-fueled engines demonstrated Brazil's technological and commercial competitiveness on an international scale.[15]

Government intervention incentivized both the commercial production and purchase of ethanol cars. Government and private interests successfully

FIGURE 1: São Paulo State Telephone Company, Telesp, became the first company in Brazil to adapt its Volkswagen service vehicles to ethanol cars over the summer of 1977.

Source: Antonio Carlos Piccino, *Acervo Globo*, June 22, 1979. https://oglobo.globo.com/econo mia/carros/cachacinha-primeiro-carro-alcool-fabricado-em-serie-chega-aos-40-em-otima -forma-23797322.

lobbied the country's four major multinational car companies (Ford, General Motors, Fiat, and Volkswagen) to produce ethanol cars exclusively for the Brazilian market with extensive market guarantees from the government in late 1979.[16] Incentives drove consumer support. Any kind of automobile remained a luxury for a wealthy few in Brazil: only 5 percent of the population owned a car by 1980.[17] As such, ethanol cars became many middle-class Brazilians' first car in the 1980s, in part thanks to the incentives Proálcool provided.[18] As sales rose and then fell, the government subsidized fuel prices and ethanol car prices below gasoline cars, reduced various ethanol car taxes, and provided longer lease terms for such cars to draw consumers to the ethanol car market.[19] These government subsidies drove up Proálcool's financial bill, but as long as ethanol prices remained low, the initiative remained popular. By 1985, ethanol accounted for over a third of the nation's car-fuel needs.[20] Ninety-six percent of all new cars sold in the country ran exclusively on ethanol, and to service this growing fleet, national ethanol production reached over 9 billion liters after the 1984–1985 harvest.[21]

Still, growing sales of ethanol-fueled cars and the continued ethanol mixture in the national fuel supply required an exponential expansion of sugar production to meet growing national demand for ethanol fuel, with dramatic environmental implications. Monocultural sugarcane production has been associated with extensive deforestation and extreme soil depletion since its rise as a global plantation crop.[22] However, Proálcool quickly pushed Brazil's sugarcane production to unprecedented levels. Nationally, the program single-handedly quadrupled cane production between 1975 and 1984. Land used for sugar cultivation nearly tripled nationally, expanding from over 3.8 million acres in 1972 to nearly 9.2 million acres in 1983. In São Paulo state, land cultivation for sugarcane expanded from 1.5 million acres to nearly 4.5 million acres, accounting for nearly half of all cane cultivated in the country.[23] Sugarcane expansion in Ribeirão Preto, São Paulo, the country's leading ethanol-producing region, directly caused the deforestation of Brazil's Atlantic Forest between 1962 and 1984.[24] Monocultural production of sugarcane also indirectly advanced deforestation by pushing other agricultural industries into new, less desirable lands. This process started with the displacement of the citrus industry in the 1980s and continued with a more aggressive, long-term pressure on land use that continues to push agro-industries into new lands to this day. Ethanol-driven demand for land pushed production of agricultural products like soy and cattle into the once undesirable Cerrado lands in the highlands of Central Brazil and eventually toward the Amazon rainforest in the north.[25] Sugarcane production expanded so quickly that the Pastoral Land Commission, a church organization, claimed in a pamphlet that the Proálcool had turned Brazil into "an ocean of sugarcane," and the borders of Brazilian agro-industries and the country's forests had changed with it.[26]

Expanded ethanol production also brought decades of extensive water pollution to sugar-producing regions. Alcohol distillation produces 10–16 liters of a liquid byproduct called vinasse for every liter of ethanol produced. For centuries, producers threw this acidic liquid byproduct in local waterways. Made of nitrogen and potassium phosphates, among other organic materials, vinasse dumping causes a catastrophic collapse in water flora and fauna as the product absorbs the oxygen in the water and asphyxiates fish, eventually rendering the water non-potable. As vinasse decimated water quality and fisheries in sugar-producing regions, public demand led to government oversight of vinasse dumping. Sugar producers and government policies instead encouraged that the nitrogen-laden byproduct be used as a fertilizer alternative. However, given its expensive storage requirements, producers often failed to comply with regulations because it was cheaper to simply dump the acidic liquid in local waterways.[27] Weak government enforcement and strong economic investment in the country's alternative energy supply hurt Brazil's waters and those that lived near them for decades. However, producers eventually remarketed this byproduct as an exportable fertilizer alternative. Still, the sugar industry remained one of the largest industrial polluters in São Paulo state in the 1990s, and vinasse spills still threaten Brazilian waters to this day.[28]

Ethanol's rapid expansion also had social costs. The expansion of production required an influx of hundreds of thousands of agricultural laborers to work the expanding sugar fields.[29] Proálcool partly succeeded because of legally structured shifts in labor relations that had begun in the 1960s and accelerated in the 1970s, which quite visibly transformed the "modern" rural labor force into one dominated by transient temporary workers. Notably, employers responded to the 1963 Rural Worker Statute (Estatuto do Trabalhador Rural, ETR), which extended urban workers' labor rights to full-time rural workers by shifting to unprotected temporary workers rather than full-time employees. In conjunction, the 1964 Land Statute (Estatuto da Terra, ET), a paltry attempt at agrarian reform under the dictatorship, empowered producers to consolidate their holdings and further displace workers.[30] This legal encouragement of large-scale land holdings and worker disenfranchisement kept large-scale sugar owners' production costs artificially but competitively low. The imposition of such conditions on workers and small landowners made the rapid expansion of an agro-industrial energy policy like Proálcool possible.

These changes, both visible and obscured, drew increasing public criticism of the program by 1983. Slowly, government officials, industry leaders, and the broader public alike began to question the basic assumptions that underwrote the intensification of ethanol production. Within government, the powerful minister of planning Delfim Neto questioned the inflationary influence of the program and froze its financial support in 1982.[31] Beyond government, some of the interest groups that had benefited most from the program began to question its long-term durability. For example, auto producers began to openly doubt sugar producers' ability to provide sufficient alcohol to meet demand.[32] At the same time, critics like outspoken University of São Paulo sociology professor Fernando Homem de Melo published attacks on the program in local newspapers and popular national magazines like *Revista Exame* and *Folha de São Paulo*. Homem de Melo notably accused sugar producers of benefiting most from the extensive financial support provided by the national program, while Proálcool's expansion exacerbated social problems; for instance, the increased demand for seasonal rural labor it created left many workers unemployed for large parts of the year.[33]

As criticism grew, avid ethanol supporter Maurilio Biagi Filho argued Proálcool's positive influence on the economy in his opinion piece published in the nationally circulated, daily newspaper *Folha de São Paulo* on May 9, 1983. Biagi Filho noted that increased ethanol production created many jobs for sugarcane workers in contrast to other industries in the recessionary national economy. He states, "In this harvest, in the region of Ribeirão Preto alone, . . . more than 11,000 new openings for jobs are available. This is without speaking of other producing regions where *usinas* [mills] and autonomous distilleries are able to absorb large contingents of manual labor while in other sectors of the economy dismissals are accumulating every day that passes." Biagi Filho's praise noted that the nation benefited from the program far more than private sugar interests did. In his article, he argued that producers were not making as much money through the program as it appeared to the public. Rather, "globally, we are losing, and the thing that keeps us on our feet is a business philosophy that reigns in the sector." At the same time, unrest brewed in the fields of Ribeirao Preto as exploitative labor practices pushed temporary workers to organize *grevinhas*, or small strikes, to demand better pay in 1983.[34] Rather than focusing on this, Biagi Filho celebrated rural laborers' new "employment benefits," like worker registration that provided health care, social security, and basic labor rights, and he claimed that sugar producers remained "in solidarity" with workers' campaigns to support the unemployed in the city. Such proclamations belittled the struggles of seasonal migrant sugarcane workers, who in fact faced job insecurity, poor pay, and violent working conditions.[35]

Instead, as doubts about the program became more pronounced during the 1980s, Biagi Filho used the promise of energy independence, which had buoyed government interests in ethanol for decades, to regenerate a wider enthusiasm for the ethanol program. He asserted "all of us Brazilians must, here and now, support and incentivize Proálcool. We ought to, indeed, defend our nationalist interests. The technology and labor are ours, we do not pay royalties [on the technology]; on the contrary, we have an international market that imports this, our technology." Biagi tried to rally support around the nationalist victory of having won the royalties to the ethanol engine patent in 1978 as a means to quiet attacks on private producers and restore enthusiasm about the program.

Nationalist rhetoric buoyed support for Proálcool through the mid-1980s, but by the end of the decade, the program's struggles were more visible. As Brazil fell deeper into economic crisis in subsequent years, both environmental strains from increased ethanol demand and financial strains from program subsidies derailed the fuel-replacement initiative. Ethanol shortages triggered by regional droughts in the late 1980s broke consumer confidence in ethanol cars, and reduced subsidies diminished ethanol's competitiveness as gasoline prices fell.[36] By the mid-1990s, ethanol cars had become relics of the previous decade and gasoline-powered cars had once more taken over the market for automobility.

However, more recently, increased focus on ethanol's comparatively lower carbon emissions as compared to petroleum has sustained government support for the fuel supplement initiative.[37] Brazil's commitment to the ethanol option opened the door for a flex-fuel engine that runs on any combination of ethanol and gasoline in the early 2000s. Today, flex cars rule Brazilian roads, providing a critical component of a diversified approach that integrates non-carbon-based fuels into the country's transportation infrastructure.[38]

Biagi Filho's positive assessment of the ethanol program's benefits and biased assessment of its shortcomings is consistent with present day assessments of alternative energy. In the search for non-carbon energy sources, proponents often focus on the positive attributes and belittle, or outright ignore, these alternative energies' broader social and environmental costs. Proálcool created numerous field jobs in sugarcane production, but it relied on an exploited, underpaid temporary labor force to fill them. Ethanol consumption promised economic savings, and later lower carbon emissions, but this same production increased environmental degradation as a result of deforestation and water pollution. Just as Biagi Filho created a narrative about sugarcane workers and failed to address Proálcool's environmental costs, policymakers today continue to focus on the energy possibilities of hydroelectric, nuclear, ethanol, solar, and wind sources without fully considering the environmental and human costs, be it land use change, population displacement, water pollution, or the exploitation of human labor that accompany large-scale production of any kind.[39]

Biofuels continued to be touted as a potential solution to global fuel demands. In 2007, the United States and Brazil, the world's two leading ethanol producers, signed an agreement to promote this alternative fuel throughout the Western Hemisphere.[40] More recently, Brazil introduced the RenovaBio program in 2017, which incentivizes ethanol production to diminish the country's transportation-related greenhouse gas emissions through tradeable carbon savings credits (branded as "CBios") to meet the ambitious goals of the 2015 Paris Climate Agreement.[41] However, biofuel's critics have focused on its expansive and intense land-use requirements. As damaging as fossil fuel use may be, some argue that the sheer amount of land necessary to accommodate such a heavy dependence on biofuels makes petroleum appear a far more economical and spatially efficient fuel.[42]

Still, technological advancements have sweetened ethanol's appeal. Sugar producers began selling electricity produced from another sugar byproduct, bagasse, the fibrous pulp left over from pressed cane, to the government in the 1980s. Use of the thermal energy produced from burning these dried husks allowed sugar and ethanol producers to use less fossil fuel in powering their mills and distilleries in a process called co-generation.[43] Scientists also adapted sugar-based ethanol into bioplastics, reducing the use of petroleum-based plastics.[44] However, these innovations helped justify continued government support, so expanding ethanol's uses helped only further silence the industry's associated costs.

Brazilian ethanol's history reminds us that alternative energy sources come with costs and that simply shifting to non-carbon fuels will not solve the world's energy situation. Moreover, despite its promotion as a solution to both climate and energy crises, ethanol fuel mixtures may even contribute to increased oil consumption. As a fuel that acts as a supplement to petroleum, ethanol has helped increase society's dependence on individual transportation, a move that accelerates problematic global consumption patterns. Ethanol fuel has become part of a history of energy "additions" that contribute to anthropogenic transformations of the world around us.[45] As policy debates about ethanol's long-term benefits continue, wider questions about the efficacy and equity of global non-carbon energy industries grow.

There is no easy fix to the world's energy issues, but parts of Biagi Filho's assertion still ring true: the Brazilian ethanol industry remains the only large-scale alternative fuel initiative to have successfully integrated into the economy on such a scale. The government's commitment to diminishing foreign energy dependence through diversified domestic energy options drove the creation of the National Ethanol Program in the crisis-riddled 1970s. At a time when climate change has made reducing carbon emissions critical to our global future, a deeper understanding of the complex development of one of the world's premiere alternative energy industries, the now-renowned Brazilian ethanol industry, offers both a roadmap and a cautionary tale for future transitions. In Brazil, the nationalistic importance of energy independence encouraged the state to continually support private producers in building an alternative fuel industry despite clear social and environmental costs. Perhaps a complete view of the economic, social, and environmental costs of this alternative fuel will help global society search for less iniquitous energy solutions in the twenty-first century.

MAURILIO BIAGI FILHO
"The Ethanol Is Ours" (1983)*

The National Ethanol Program (Proálcool) is effectively experiencing a new period, thanks to the space that the mainstream press has been giving to the discussion of the problems related to it. This is healthy, as the debate always searches for new alternatives. And it is this that we all want.

The validity of this program is no longer questioned, since every Brazilian feels proud and responsible knowing that here, in our country, with technology and manual labor of Brazilians, we were able to successfully develop the largest and most important renewable energy project in the world.

In the beginning of this harvest, in the region of Ribeirão Preto alone, [which is] responsible for the production of more than 33% of national ethanol, more than 11,000 new jobs are being offered. This [is] without mentioning the other producing regions, where the *usinas* [mills] and autonomous distilleries are able to absorb large contingents of labor, while in other sectors of the economy layoffs increase with every passing day.

Now the situation for producers is worrying. At first sight, many may think that we are accumulating fortunes. This is not true, and anyone can verify the situation that we endure, upon analyzing the balance sheets that mills and distilleries keep publishing in the news. Globally, we are losing, and the thing that keeps us on our feet is the business philosophy that reigns in the sector.

We believe in what we are doing. And we know that we are making a huge social investment, as we move forward with this, the only renewable and viable alternative solution in the world. Many businessmen in the sector seek with great effort and sacrifice money through loans, guaranteed by their own family, placing the patrimony that represents years of sacrifice as a guarantee of these payments.

---

*Maurilio Biagi Filho, "O álcool é nosso," *Folha de São Paulo*, May 9, 1983, 36. Translated by Jennifer Eaglin. Reprinted with permission from Folhapress, São Paulo.

"The Ethanol Is Ours" (1983)

It is important to stress that these loans, in cruzeiros with the rarest exceptions, end up being paid on time. The situation is different from that mentioned a few days ago by Joelmir Beting, here in the "Folha" [newspaper], when he said that "the moment that Fiat of Turin sends dollars for Fiat of Betim [in Brazil], it is Delfim [Neto, Brazilian Minister of Planning] who ought to pay, instead it is 125 million Brazilians who pay."

Today, our rural worker, called pejoratively the *bóia-fria*, is experiencing new days, with employment benefits, such as registration with a professional card (which assures him all labor rights), with additional benefits thanks to the application of funds already calculated in the cost of the selling price of sugar (1%) and the price of ethanol (2%). In the Ribeirão Preto region, we are keeping up with a solidarity movement capable of raising anyone's awareness, with rural workers carrying out campaigns to collect foodstuffs, medicine, warm clothes and even money for the unemployed in the city. The mills even opened vacancies for unemployed metallurgists to harvest beans.

That ethanol is a better fuel than gasoline to move our fleet of more than a million automobiles (estimated up to the end of the year [1983]), this the whole world already knows. As it also knows that all the cars run on gasoline carry in their tank and carburetor a mixture of 20% of anhydrous ethanol. Those who already use ethanol fueled cars feel their advantage in their own pockets.

Now comes the time of light transport, of semi heavy transport and of the use of ethanol in agriculture. At the Usina Santa Elisa in Sertãozinho, the results with the 300 Volkswagen trucks run on ethanol prove that their cost is less than that of gasoline (the cost per ton/km of diesel is Cr\$31,38 versus Cr\$27,22 for ethanol). In the same mill, we already have tractors driving with our green and yellow fuel [Brazil's national colors], with exceptional results. In Ribeirão Preto, in the coming days, we will drive a bus also powered by ethanol in yet another pioneering project.

Many blame the government for the failures of our economy. But recently, in Brasilia, during the third National Meeting of Autonomous Distilleries, promoted by the Society of Producers of Sugar and Alcohol (Sopral), we heard from our Minister of Industry and Commerce Camilo Pena the following words: "Our ethanol is competitive with petroleum. And, even if it weren't, our choice ought to be for it, because Brazil does not have dollars to import petroleum."

To this statement, we also add that made in that producers' meeting, by commercial director of Petrobras, Carlos Sant'Ana: "Today, ethanol costs Petrobras less than petroleum." General Oziel de Almeida, president of the National Council of Petroleum, guarantees also that "the ethanol bill is a surplus" even as the ex-minister Calmon de Sá warns that "to leave leftover cane on the floor is stupid," suggesting that the highest amount possible of ethanol be produced.

The truth is that the entire Nation feels the importance of Proálcool, the Brazilian government's most successful program. Today, the sugar sector manages a contingent [labor force] of 1.5 million Brazilians, who work directly in the sector, managing a Gross National Product of around US$15 billion. This year alone, Proálcool will represent a savings of US$1.5 billion in foreign exchange, and since its creation has brought a savings of more than US$6 billion.

If much needs to be done still, it is necessary, here and now, that all of us Brazilians, we know to support and incentivize Proálcool. We ought to, yes, defend our nationalist interests. The technology and the manual labor are ours, we do not pay royalties, on the contrary, we have an international market to import this, our technology.

2 "Coal Will Be the Primary Fuel of the Future"

Yoshimura Manji on the "Fuel Question"

VICTOR SEOW

IN JAPAN, as in many other countries, the period after World War I encouraged increased anxieties about development and access to necessary resources. Globally, the preceding years of fighting had been marked not only by a devastating loss of life but also by material shortages that had resulted from shipping blockades and protracted combat. Among leaders of postwar societies, self-sufficiency became an especially prized objective, and scarcity a particularly feared fate. Within this wider context, many Japanese commentators started expressing concern over their country's present and projected lack of energy resources, a concern that found expression in what they termed the "fuel question" (燃料問題).[1]

At the time, the Japanese empire consisted of the four large islands of Honshu, Kyushu, Shikoku, and Hokkaido ("Japan proper"); hundreds of smaller islands; its colonies of Taiwan, Karafuto, and Korea; its mandate in the South Seas; and its leasehold in southern Manchuria and the railway zone that extended northward from it. In spite of this territorial expanse and the access to resources it afforded, a persistent refrain in Japanese public discourse was that Japan (by which it was meant Japan proper) was a small island nation lacking natural endowments—a sentiment that would be popularized in the 1930s by its characterization as a "have-not country" (持たざる國), a place where geographic and geological circumstances were believed to have created a comparative developmental disadvantage.[2] The fuel question both reflected and contributed to the endurance of this widespread belief.[3]

There had been earlier fears about access to fuel. For instance, in the lead-up to the Russo-Japanese War (1904–1905), Japanese naval planners worried about the coal that would power their fleet. There were sizeable coalfields in the Japanese home islands, particularly in Kyushu to the south and Hokkaido to the north. But most of the coal that was mined was not well suited for fueling battleships. Its calorific value was too low. Worse still, it gave off tactically compromising black smoke when burnt. Such concerns were allayed only after the navy managed to secure additional imports of better-quality coal from Britain.[4]

What made the Japanese interwar fuel question different from such prior expressions of anxieties over fuel, though, was that it was not so much about overcoming a specific challenge, such as obtaining sufficient stocks of naval fuel in anticipation of a coming war. It was, rather, more about confronting a perceived deficiency that had implications for the nation on multiple fronts. As Okunaka Kōzō, then secretary of the Fuel Society of Japan, wrote in 1927: "The question is really one of vital importance in that its satisfactory solution is closely interwoven with all issues of national strength, defense, and industrial development, in short the whole system of national well-being."[5]

A key facet of the fuel question was Japan's lack of domestic oil. The country had limited reserves, primarily around Niigata, but, largely on account of cheap imports, Japanese producers ended up focusing less on developing domestic reserves and more on refining foreign crude.[6] And if the preceding war had one lesson to impart, it was that oil had now become a key determinant of the age's geopolitical winners and the losers. In powering mechanized means of transport from tanks and trucks to ships, submarines, and airplanes, oil helped redefine the modern battlefield and the very nature of war. Oil-burning ships, for one thing, were faster, easier to fuel, and emitted less smoke. The Imperial Japanese Navy, which first installed oil-fired boilers on one of its destroyers in 1915, began, after the war, an over-a-decade-long process of converting its entire fleet to run on oil.[7] Among the many Japanese who began touting the newfound importance of oil to national defense, it was common to quote for effect or emphasis French prime minister Georges Clemenceau's famous quip that "a drop of oil is worth a drop of blood."[8]

Beyond the battlefield, oil was also being consumed in increasing quantities for commercial and industrial purposes, driving engines across a range of vehicles and an assortment of machines. Nevertheless, in public assertions as to why Japan needed access to more of this resource, its military uses were paramount.[9]

The post–World War I period also saw Japan relying on greater quantities of foreign coal. As in the case of battleship fuel, this was mostly related to a mismatch between the types of coal found in domestic deposits and the types of coal in demand. Most notably, Japan had very little of the type of coal that could be coked to make steel, a vital companion in the age of industrial manufacturing. The increase in imports, which came primarily from China and Indochina, was also a byproduct of difficulties the Japanese coal mining industry faced with high production costs and low worker efficiency, the latter of which had been blamed on the thinness of Japanese coal seams, a geological particularity which made the use of mechanical tools more challenging. In 1923, Japan became a net importer of coal, a status it would largely maintain in the following years.[10]

These developments alone may not have been an issue. They presented a problem primarily because they ran counter to the ideal of self-sufficiency that had become widely embraced in Japan, as elsewhere—which was itself mainly a result of the military having experienced strategically compromising material shortages in the preceding war.[11] Just when Japan's leaders started championing resource autarky, its military and industries were increasingly more dependent on foreign sources for the fuel. This growing dependency on energy imports, coupled with a domestic lack, was at the core of the fuel question.

Among the responses that the fuel question drew, few would be as significant as the founding of the Fuel Society of Japan in 1922. Two years prior, the Ministry of Agriculture and Commerce had set up in Saitama, just to the north of Tokyo, the Fuel Research Institute. Like the fuel boards and commissions springing up in other countries after the war, the purpose of this new governmental outfit was to bring together researchers to study how fuels might be used more efficiently given escalating fears over their short supply. Soon after the Fuel Research Institute was established, its head, mining engineer Yoshimura Manji, began meeting with several fellow researchers and others who were interested in fuel matters to talk about specialized topics such as the low-temperature carbonization of coal. This Fuel Discussion Society, as they called themselves, would become the basis of the Fuel Society, as those who took part in the earlier ad hoc gathering decided to pursue a more formalized arrangement. At its founding, the Fuel Society consisted of twenty-three members; within three years, its membership reached 1,500.[12]

The Fuel Society, whose ranks included scientists and engineers like Yoshimura as well as industrialists, military officials, and government bureaucrats, continued serving as a site for the exchange of ideas and information across these sectors. But its members were, from the very beginning, determined to not just keep the conversation among themselves. They actively worked to bring the fuel question to the public's attention too. As one member described it: "When a moving train is about to cross over a broken bridge, it is incumbent upon those passing by, as people near to the train, to endeavor to save it. As those who are close to our country's fuel question, we should alert our compatriots to the current danger and work to mitigate the fuel situation for the sake of our national livelihood."[13]

In attempting to raise public concern, the Fuel Society undertook measures on several fronts. It organized various events from monthly public lectures to annual conferences. These featured experts who spoke on a range of topics that generally related to either recent technological innovations and other advancements in the world of fuel research or the nature of the fuel question in Japan and beyond. Many flocked to these talks. The event that drew greatest number of people, though, was a fuel exhibition that the society held in Tokyo in 1930. At this exhibition, visitors were treated to extensive displays of new products from alternative liquid fuels to coal-saving appliances and were fed a wide assortment of facts about fuels through posters, write-ups, lectures, and short films. This went on for ten days, attracting almost 40,000 visitors.[14]

The Fuel Society also tried to engage a wider audience through print, publishing, to that end, a monthly periodical, *The Journal of the Fuel Society* (燃料協會誌). In this journal's pages, one could read the transcripts of many of the talks and discussions the society organized as well as news, reports, and editorials about developments in fuel research, the state of Japan's energy economy, and the society's various activities. In 1925, the society ran an essay contest inviting entries that identified important issues in regard to the fuel question and best conveyed these to the general public. The winning essay, which, among other things, attempted to provide an "ethical" justification for the use of force to secure fuel rights, was published in the journal the following year.[15]

The Fuel Society would be active on several other fronts. A notable initiative involved continuously petitioning the government to introduce a national fuel policy.[16] In addition, many of its members had, within their respective spheres of academia, industry, government, and the military, also helped shape in ways big and small the course of Japan's energy sector. However, as a society, its most significant achievement was likely to make, as it had set out to do from the start, the fuel question an issue of widespread concern. Although different actors would propose different solutions to problems of fuel access, the basic premise that Japan was in or approaching an energy crisis was something that remained essentially unchallenged.

Yoshimura Manji, the founding president of the Fuel Society, had studied mining and metallurgy at Tokyo Imperial University, graduating in 1906. Between 1913 and 1915, several years before the government appointed him to lead its new Fuel Research Institute, he studied in Europe and the United States. From this stint abroad, he seems to have internalized two main insights. The first was that Japan was poor in fuel resources, especially as compared with Britain and the United States, which held great mineral wealth within their territorial boundaries and across their many colonies. The second was that policymakers needed to foster an awareness of the fuel question among the Japanese public and make it an imperative to conserve resources and to subject them to efficient use.[17]

The accompanying essay by Yoshimura in this chapter, "An Overview of Our Country's Fuel Question," appeared as the opening piece in the very first issue of the *Journal of the Fuel Society*, published in August 1922. In it, Yoshimura defines the perimeters of the fuel question and makes a case for the widespread implications it had for Japan's present and future. He begins by laying out the global transformations that gave rise to the fuel question. He then proceeds to address, in turn, the "coal question" (石炭問題), the "liquid fuel question" (液體燃料問題), the "household-use fuel question" (家庭用燃料問題), and finally the "new fuels question" (新燃料問題). As much as the fuel question was presented as an overarching issue of general concern, it was also understood as a range of challenges tied to the specificities of fuel type. By pulling together these threads, the essay offers a fascinating set of reflections on the intertwined relationship between energy and society against the backdrop of rising resource anxieties.

To begin with, Yoshimura's account of the emergence of the fuel question is striking. He recognizes that one may have been prompted to pay more attention to fuels due to a perceived "desirability of instructing against the waste of natural resources and endeavoring toward their economical use." As the essay suggests, such a response drew inspiration from both the American conservationist movement and other fuel-efficiency initiatives in Britain. Or one may have been moved to focus on fuels, Yoshimura proposes, because of strategic calculations that foregrounded a "need to investigate the means of securing fuel supply for warships and other such vessels." This would certainly be consistent with the post–World War I surge in anxieties over access to petroleum resources. Interestingly, though, the most important dimension to consider, at least to Yoshimura, was the economic one.

This is interesting not just because of how he places economic imperatives over military demands when the latter had been to many others the primary driver for the rise of the fuel question in Japan, as mentioned earlier. It is also interesting because of how exactly he delineates this particularly critical "economic dimension." Rather than framing this dimension in terms of the needs of the national economy—the "wealth" in the "wealth and power" that commonly defined nationalist pursuits in that era—he focuses, instead, on the "masses" whose "material necessities" were supplied by various "economic institutions," the industries and infrastructures under the "systems of modern production."[18]

Yoshimura's focus on the masses stemmed from his contention that the most pressing issue of the time was the "awakening of many classes of people." Recent episodes of collective action in Britain and the United States were, to him, evidence of the "mass's self-emergence." By this he meant that while non-elite people had been mostly occupied with "simply maintaining a livelihood and accomplishing survival" in the past, they had come to desire more, seeking to "develop one's individuality and assert oneself." And this was, Yoshimura suggests, a positive thing. Unlike countless other elite actors here and elsewhere who criticized labor movements as disruptive to the social order or damaging to the national economy, he appears to have regarded them as justified in their striving toward "cultural refinement" and "spiritual development."

The problem, as he saw it, was that if this kind of striving were to be supported, it would require a material foundation. And this was where the fuel question came in, for such a material foundation was inseparable from the sources of energy that powered it, ever more of which were being demanded. Yoshimura singles out two factors as having led to such an outcome. The first was Japan's increase in population; the second, the improvement in livelihoods.

Population increase was a topic much discussed by Japanese social commentators in the interwar period. Although there had been earlier concerns over the possible negative social effects of population expansion, the 1920s witnessed a heightened sense of crisis as the increased availability of population statistics and, hence, evidence of population growth coincided with the post–World War I economic recession.[19] For Yoshimura, as it would be for many others who weighed in on this issue, the problem here was one of constrained resource availability, with more rice needing to be cultivated and more residences built to feed and house the swelling population.

Moreover, these challenges were compounded, he contends, by a general improvement in material well-being, as "everyday necessities now arrive in a welter of variety and in great number." Mass production and advances in transport had placed into the hands of a greater number of people a widening array of goods to consume. Once more, to his credit, Yoshimura did not begrudge the fact that items that may have once been for only the few were now available to the many. But to support increased demands for newfound necessities, alongside those for food and shelter, "systems of modern production have grown progressively vast" and these all "require large sources of power and consume fuel in great amounts." This was, to him, the essence of the fuel question.

Having laid down the stakes, Yoshimura then claims that "coal will be the primary fuel of the future." In claiming this, he directly cites British mining engineer John Cadman, who had recently made this assertion. This claim is striking because, later in the piece, Yoshimura also states that "the age of solid fuels is passing, and the world is entering the age of liquid fuels."[20] Suspending for a moment any consideration of this seeming contradiction, we can see that his reasoning rests primarily on the comparative bounty of coal relative to alternative sources of energy. In the case of petroleum, the world's largest deposits of which had not yet been discovered, Yoshimura notes that the amount of this liquid fuel produced in the past few years, when converted to its equivalent in coal, was but a fraction of the actual amount of coal extracted over the same period. Under these conditions, and with an acknowledgment that the potential of future oil and coal fields remained uncertain, he concludes that in regard to coal "no comparable source can be discerned."

Precisely because coal was so important, its use required prudence. Yoshimura points out how a considerable amount of coal was typically lost between the coal seam and the point of final consumption, something particularly regrettable because there was "no hope of its regeneration." He then outlines the measures by which coal might be efficiently used. In this, he stresses the need to be mindful of the specific properties of different types of coal and the uses to which they can be correspondingly put. Most notably, he suggests that coal should, whenever possible, be processed through dry distillation into coke, gas, or oil. Such transformations in form would allow the fuel to be put to more optimal use in ways otherwise impossible if it were burned directly. Historians have drawn attention to the persistence of earlier kinds of energy in periods defined by their supposed successors, such as muscle power in the age of coal or coal in the age of oil.[21] Here this persistence was further facilitated by reworking coal to derive other fuels to power processes and devices that coal in its original state could not have powered.

Throughout his essay, Yoshimura situates Japan and its fuel question within a wider, transnational frame. He makes repeated reference, in particular, to coal's production and consumption in Britain and the United States. Yoshimura notes, for instance, long-standing concerns over coal in both countries, referencing British political economist William Stanley Jevons's famous 1865 book *The Coal Question*, which had sought to quantify the limits of Britain's coal-dependent hegemony.[22] What Yoshimura seems to have been particularly keen on emphasizing was the extent to which the government of each had committed to fuel research, something that further intensified after World War I. Similar efforts in Japan, including the establishment of the aforementioned Fuel Research Institute that he himself headed, could be understood (and, insofar as he needed to make a case for them, justified) in this context.

Comparison can be a double-edged sword. For Yoshimura, as it would be for many other Japanese commentators, placing Japan alongside other countries deepened a sense of national inadequacy. Citing coal deposit figures for the United States, Canada, Britain, Germany, Australia, China, and Japan, as had been recently published by the International Geological Congress, he highlights how "Japan possesses only around a five-hundredth part of the United States' deposits and a hundredth of China's" and how "it is easy to foresee that Japan will run out of coal much faster than the United States,

China, and the other countries mentioned." Were coal indeed the "primary fuel of the future," Japan's future did not seem especially bright.

In our current historical moment, just over a century after Yoshimura's essay of 1922, many of the anxieties around energy expressed in it persist both in Japan and across the wider world of which it is part. Japan remains a fossil-fueled power. As of 2022, it is the world's third largest coal importer and sixth largest oil consumer.[23] Japan's concerns over access to energy resources have been particularly pronounced in the decade following the 2011 Fukushima Daiichi nuclear disaster, in which new coal-fired power plants have been built even as formerly disconnected nuclear reactors have been placed back online.[24] Still, in Japan, as in other comparatively more industrialized societies, the more pressing issue may be less about securing cheap and abundant sources of energy to maintain high levels of consumption than about confronting the climatic and ecological devastation wrought by the profligate consumption patterns of the preceding century.

YOSHIMURA MANJI
"An Overview of the Fuel Question in Japan" (1922)*

THE ORIGINS OF THE FUEL QUESTION
The fuel question may be considered in ethical terms as the desirability of instructing against the waste of natural resources and endeavoring towards their economical use, or, from a military perspective, as the need to investigate the means of securing fuel supply for warships and other such vessels;

---

*Yoshimura Manji "An Overview of the Fuel Question in Japan," *Journal of the Fuel Society of Japan*, 1, no. 1 (1922), 1–9 (slightly abridged). Translated by Shi-Lin Loh. The front matter specifies that this piece is an editorial and that Yoshimura is a member of the Fuel Society. In the title and throughout the text, Japan is referred to as 我国 (*wagakuni*, lit. "our country"), which is a standard term that often carries nationalistic overtones; for simplicity's sake it is rendered in this translation as *Japan*.

yet its economic dimension should be viewed as the most crucial. Many economic institutions of the modern age provide the masses with material necessities, allowing them physical health, latitude in their livelihoods, and the opportunity for quiet contemplation, thus securely orienting them towards the path of cultural refinement.

A distinctive feature of the present era and its most conspicuous fact is the awakening of many classes of people, and the force of these masses' self-emergence has become most prominent following the end of the great war in Europe, one part of which can now be observed in the labor movements of the various countries. The recent strikes by coal miners and railroad workers in England, as well as the trade union activism in America, provide us with fresh examples of this phenomenon. Even though these movements in one sense disrupt the order of their countries and impose economic damages, I venture to ask what in the first place is causing them, and in my view, they are based on the masses' self-emergence and awakening; whereas they previously stopped at simply maintaining a livelihood and accomplishing survival, today they are not satisfied with these alone, and they make headway with asserting themselves and developing their character, manifesting the sentiments of seeking to stand equally with others and contribute to the progress of worldly affairs. If they merely sought to maintain one's own existence, it would be a simple affair; yet in order to develop one's individuality and assert oneself, there must be the latitude to do so in one's livelihood, which must also extend to the temporal and material dimensions. Without this, one cannot expect to achieve improvement and progress. As expressed by the great English scholar, Prof. Marshall, a life without latitude holds no chances to gain friends, and is equipped with no path for cultivating character, no leisure for quiet sitting or self-enjoyment; spring comes and the birds twitter, but one sings not, the flowers bloom, but one smiles not, and is unable to nurture one's noble intellect; the greatest reason for all these is, in truth, poverty.* That is why, for the sake of granting spiritual development to all classes in society, it will first

---

*Most likely a reference to the economist Alfred Marshall (1842–1924). This sentence is supposed to be a quotation, but no source is provided.

"An Overview of the Fuel Question in Japan" (1922)

be necessary to establish a foundation that is material. With regard to this demand, two detrimental obstacles may be raised. The first of these is the increase in the population of humankind and the second is the improvement of livelihoods. The increase in the human population, as collected in statistics for the decade from 1900 to 1912 [sic], shows that for every 1,000 people 120 more are added on average throughout the world. In other words, every 10 years the population increases by 12 percent. Now, supposing that every 10 years it grows by about 10 percent, in roughly 17 years it would have doubled. According to sources in England which have conducted research and calculations, that country's population, which numbered 40,830,000 in 1911, is forecast to grow to 81,260,000 in 2051. In short, it is positioned to more or less double in 140 years. The rate of population growth in Japan in the aforementioned decade, taken as 137 people per thousand, works out to a doubling of the population in about 50 years. Recent annual statistics show that Japan adds 600,000 people each year, and thus, if the current population of the city of Tokyo is assumed to be 2,050,000, the increase in people will surely swell this metropolis in 4 years' time. Moreover, if each person consumes on average one *koku** of rice each year, this would necessitate an increased harvest of 600,000 *koku* every year, and to enable that, 30,000 hectares of paddy fields would have to be newly cleared. Furthermore, housing this increased population would require building at least 100,000 new residences.

With regard to the improvement of livelihoods, along with clothing, food, and shelter, old notions have been rectified, and everyday necessities now arrive in a welter of variety and in great number; aside from demanding the bountiful harvest of rice and other staple grains, as well as of cotton yarn and sheep's wool and steel and other general manufactures, whose variety and number have also increased excessively. Accordingly, production in the present era occurs on a large scale, and transport has also grown fleet.

---

*Approximately 140–150 kilograms, though originally a measure of capacity. For a detailed explanation, see the National Diet Library's reference Q&A, last modified January 22, 2011, https://crd.ndl.go.jp/reference/modules/d3ndlcrdentry/index.php?page=ref_view&id=1000076982.

Yoshimura Manji

[Yoshimura goes on to vividly describe the progress in the production of goods and the speed and power of transport.]

It is in striving to respond to such an increase in human population and the improvement of livelihoods that the systems of modern production have grown progressively vast. General industries, like steelmaking, have reached a large scale that leads to frantic production day and night, ferrying out goods in an attempt to balance out successive demands from East and West, which also speeds up the transport of these goods; led by railroads and steamships, automobiles and airplanes have subsequently arrived and acquired distinct functions in their uses. Every one of these things—general factories, the steel industry, railroads, steamships, airplanes, automobiles, and the like—all require large sources of power, and consume fuel in great amounts, while their numbers grow ever greater with the years. Here we now see the towering rise of the coal question, the liquid fuel question, the gas question, the new fuel question, and other issues related to the general question of fuel.

THE COAL QUESTION

Amongst the many problems to do with fuel, that which occupies the most crucial position, and which Japan in particular must consider deeply, is the problem of coal. Although other sources of power and fuel exist, such as hydropower and petroleum, their importance lags far behind that of coal. World hydropower reserves available can generate electricity at 440,101,000 horsepower, and have an actual electrical output of 22,832,350 horsepower. Japan's share of both these amounts is approximately 7,000,000 and 1,200,000 horsepower respectively; converted to coal, the first amount is equivalent to 42,000,000 tons, while the second amount is 7,200,000 tons. Put another way, if we take Japan's current hydroelectric capacity in terms of coal, we would need 7,200,000 tons of this fuel. In the event that the entirety of Japan's available electrical capacity is exhausted, it would be used up as a supplement for this 42,000,000 tons of coal. And if we are to keep ensuring that Japan does not consume more than forty million tons of coal, even ignoring the impossibility of relying on hydropower as a future source of power, the amount of coal consumed continues to increase year after year, with no prospect of reaching a stable state. Here I omit the question of petroleum [. . .]. I

"An Overview of the Fuel Question in Japan" (1922)

will only mention that [. . .] petroleum at present is no more than a thirteenth or fourteenth of coal; as for deposits of coal and petroleum and their respective durations, although these are not easily demonstrated in figures, it is broadly possible to discern that the deposits and duration thereof for coal appear to be far larger than those of petroleum.

As Sir John Cadman* of England has remarked to an academic association in that country, while even today much about petroleum fields remains mysterious, these unknowns belong to the same category as undeveloped coalfields, and to extrapolate from present conditions, the mainstay of future fuels will be coal, in the absence of which no comparable source can be discerned.

Coal is used in factories, railroads, steamships, and in other ways too numerous to list. Its consumption keeps growing along with civilization's progress; to take one example of this, from the United States, the per capita consumption of coal in that country is as follows:

1870	0.96 tons
1880	1.40 tons
1890	2.30 tons
1900	3.20 tons
1915	5.50 tons
1920	6.20 tons

To further emphasize the remarkableness of this trend, the amount of coal mined in Japan is as follows:

Meiji 15 [1882]	944,113 tons
Meiji 25 [1892]	3,199,763 tons
Meiji 35 [1902]	9,701,682 tons
Meiji 45 [1912]	13,716,488 tons
Taishō 9 [1920]	29,423,689 tons

---

*Baron, engineer, and petroleum technologist (1877–1941).

Yoshimura Manji

Meanwhile, regarding the amount of coal consumed in Japan, according to the estimated calculations of Ishiwata Nobutarō,* this will total 25,000,000 tons in Taishō 7 [1918], 50,000,000 tons in Taishō 17, 85,000,000 tons in Taishō 27, 125,000,000 tons in Taishō 37, and 191,250,000 tons in Taishō 47; in other words, by Taishō 50 this will have reached 2 hundred million tons.† When that time arrives the population will be seven or eight times larger than at present, around 90,000,000, and each of these people will consume two tons of coal or thereabouts per year.

With regard to this increase in the amount of coal consumed, we must not fail to consider the amount of coal that remains. According to the findings presented at the International Geological Congress of 1913, the total sum of coal in the world stands at 7,397,553,000,000 tons, of which the United States holds 3,838,700,000,000 tons, Canada 1,234,400,000,000 tons, England 189,500,000,000 tons, Germany 23,400,000,000 tons, Australia 165,572,000,000 tons, China 995,587,000,000 tons, and Japan 8,000,000,000 tons, which shows that Japan possesses only around a five-hundredth part of the United States' deposits and a hundredth of China's.

How many years it will take for coal to be exhausted, and other questions concerning its lifespan are in the first place not easily resolved, but although this is not readily demonstrable in figures, if current trends in coal consumption remain the same, it is easy to foresee that Japan will run out of coal much faster than the United States, China, and the other countries mentioned. Even if that turns out not to be the case, it is still clear that a period of extreme hardship or similar distress is constantly approaching. Meanwhile, the issue of the lifespan of coal deposits further raises two or three other questions, such as the depths at which it is possible to extract coal, and the required thickness of coal-bearing strata. At present the former's extent is

---

*石渡信太郎; a reputed mining engineer and former member of the Meiji Mining Company (1875–1957).

†Taishō 17, 27, 37, 47, and 50 are nonexistent dates that the author anticipates. The Taishō emperor died in Taishō 15 [1926], four years after the publication of this editorial, and the Shōwa era started on December 25 of the same year with Hirohito's assumption of the imperial throne.

"An Overview of the Fuel Question in Japan" (1922)

4,000 *shaku** below ground surface, and the latter must be at least one *shaku*; nonetheless, technology will make progress hereafter, such that pit temperatures can be reduced and allow work to be performed even at 6,000 *shaku*, while machines will be invented that permit the economical extraction of coal from strata thinner than one *shaku*; if so, the amount of coal mined will be far greater, and the lifespan of its deposits can be extended.

Given coal's limited reserves, using it causes the loss of its form and the fading of its effects, with no hope of its regeneration; regarding this, the continuing increase in its consumption with each year gives rise to the necessity of considering mining laws, transportation and conveyance, and consumption methods. For Japan in particular, which has a mere hundred tons of coal deposits, this necessity is of the greatest relevance. If no plans are instituted at the present time, this will surely lead to an eternity of regret.

Regarding the amount of coal lost during its handling, an investigation based on the figures from the United States Bureau of Mines shows that for the extraction of one short ton (i.e., 2,000 pounds) of coal below ground, an initial 600 pounds are lost in coal pillars, coal walls and other such places that cannot be extracted.† Another 31 pounds are claimed when heaving up the coal, 82 for transportation, 13 more naturally lost in the course of transportation, 446 escaping with the smoke from steam boilers, 51 each to radiant loss and ashes, and 650 to the exhaust from steam engines, so that finally what we can use as mechanical power out of 2,000 pounds of coal is in fact only 76 pounds. In other words, this is just under 4%, and considering this fact, along with urging the improvement of laws on coal mining and handling, to broadly extend such concerns to the law on coal burning, when burning coal (and in cases where it is converted to gas and burned as such), given that the methods of obtaining steam are extremely lax, the conversion of steam engines, the use of steam turbines and gas engines, and switching to water as an ideal medium are some other methods that deserve thought and research.

---

*A *shaku* is a Japanese unit of measurement, approximately equal to 30.3 cm.

†炭柱 and 炭壁 are the original terms, rendered here as "coal pillar" and "coal wall" as apparent references to mining methods.

There are on the whole four methods for using thermal power from coal to derive motive power, as follows:

1. Burning coal as it is using hand shovels or stokers
2. Burning coal compressed into briquettes
3. Burning coal that has been pulverized
4. First converting coal to gas and burning the gas, or using a gas engine

Which of these methods is most efficient and well-suited to our purposes depends on the variety of coal being employed; generalizations should be avoided. [. . .]

As with all unprocessed materials like fresh fruit and raw fish, coal will soon change its nature if left in its original state and grow liable to disintegrate into powder form. Or, depending on its variety, it might spontaneously ignite, and therefore it is best to be able to use coal right away in the place where it is mined, for otherwise it would be impossible either to convert and convey it as gas, as well as to turn it into electric power. To go a step further, coal should be subjected to the dry distillation method and triply converted into gas, petroleums, and coke, as these three forms could be ideally put to various uses according to their respective characteristics. Here we must mention how coke, especially, does not change its makeup even with the passing of time and is extremely stable; in terms of coal consumption, this means not only that minimal loss can be anticipated, but also that a decrease in the quality of coal can be avoided.

That the coal problem occupies a crucial place in a country's economy requires no further explanation, which is why in Europe and the United States research on it is already unstinting, with its progress particularly visible in England and the United States. In what follows I will summarize the reasons for this while making occasional observations on the present state of Japan's fuel industry. In 1865, Professor Jevons wrote a book called *The Coal Question* in England, which discusses how deeply coal is intertwined with the rise and fall of a country's fortunes.* At that time, coal mining companies

---

*William Stanley Jevons, *The Coal Question; An Inquiry Concerning the Progress of the Nation, and the Probable Exhaustion of Our Coal Mines* (London: Macmillan, 1865).

"An Overview of the Fuel Question in Japan" (1922)

only searched for coal in strata that were easily accessed, shunning places that were hard to reach, but based on this former situation, if the costs of digging did not increase, coal would soon become unavailable for mining; this same professor postulated that, at this rate, England would promptly fall into the plight of becoming an inferior straggler compared to other nations, and warned his fellow people about the risk of this happening.

To begin with, England's might is the result of a dual advantage relating to coal in its industries: due to coal's inexpensiveness, it has been able to manufacture pig iron at low prices and also various machines in the same manner; moreover, the inexpensiveness of coal and machines also implies that manufactured goods will be affordable, and further that coal's affordability will bring about favorable results in the promotion of foreign trade. In short, most imports coming into England are articles of raw materials in bulky states which, on repatriation, fill the hull of a ship to its brim due to the [overseas] desire for non-bulky English manufactures, and in these cases, ships return laden with [the products of inexpensive] coal, which in turn cannot but make their freight charges inexpensive. Ultimately, the English depend on affordable raw materials from foreign countries to enjoy the convenience of affordable daily necessities, which is a key dimension of England's present prosperity. Furthermore, the author of *The Coal Question* indicates that coal is the cheapest source of motive power; though the powers of water, tides, wind, and sun may be directly harnessed, in the end these cannot be replacements for coal. We may say that coal's true value, in tandem with the progress of the world, grows ever more apparent.

In 1871, England's government appointed commissioners to research coal and to conduct a detailed investigation into issues including the country's available coal deposits, how much could be extracted and how long before they were exhausted.* In 1904, a second similar commission was established and, in addition to the amount of coal deposits, was charged with researching the handling of extraction methods and consumption methods and comparing these findings to the situation in Germany, in an attempt to derive the lifespan of their country's coal supplies by seeing how long its neighbor's own

---

*This commission was formed in response to Jevon's *Coal Question*.

would take to run out.* In the great European war, England started administering its coal by such actions as limiting prices for extraction and distribution as well as restricting exports, but it found the supply of petroleum most difficult to manage. For this reason, to provision its home territory with liquid fuel, it experimented with test drills of oil wells, and further established a Fuel Research Board chaired by Sir George Beilby,† taking pains to derive liquid fuels solely from coal as a raw material. In the United States, with regard to the conservation of natural resources, one J. W. Powell authored a report circa the 1880s on "Lands of the Arid Region," which analyzed the devastation of forests and noted the dire situation in Spain and Italy thereof as a warning to the general public.‡ Gifford Pinchot also argued for the protection of forests and asserted the use and conservation of hydropower and rivers.§ As a result of such arguments, President Roosevelt created the Inland Waterways Commission [in 1907], and the findings of its investigations soon led to its recommendations for development and conservation including not only forests but also sites of mineral deposits and other resources, on which subject a Conference of Governors was convened in Washington in May 1908. In attendance were senators from both houses of Congress, state governors, representatives of organizations, entrepreneurs, experts, and other parties, who took turns to speak on the urgency of conserving natural resources in relation to their various concerns, leading [Charles R.] Van Hise¶ to author a treatise on the conservation of the United States' natural resources as a whole.

From the start, even though the United States enjoys a vast territory and an abundance of natural resources, traits which distinguish it from

---

*For titles of the relevant British Royal Commission reports, see Keith Ramsey, *The British Coal Industry* (Bristol Branch of the Historical Association, 2003).

†A noted British industrial chemist (1850–1924).

‡Actually 1879; see *Report on the Lands of the Arid Region of the United States*, 2nd ed. (Washington, D.C.: Government Printing Office, 1879), https://pubs.usgs.gov/unnumbered/70039240/report.pdf.

§First head of the United States Forest Service (1865–1946).

¶A noted American geologist (1857–1918).

"An Overview of the Fuel Question in Japan" (1922)

England and the like, its Geological Survey [USGS] and Bureau of Mines have spent many years researching these kinds of issues, and they are joined by the non-governmental organizations of state universities and for-profit companies, not to mention private individuals who have eagerly engaged in such research. Like England, it also started to administer its coal during the great war in Europe by instituting restrictions on its distribution, price, and export. The administration of coal took up a large amount of labor and costs, and suffered criticism that doing so only imposed pointless hindrances; nonetheless, even after the war many asserted that the distribution of coal alone required the establishment of a system. This system, according to the national character of the United States, would not be favorably run by the government, but rather more dependably outside of it, and today amongst coal vendors such a system for exporting coal has been established. Although hitherto in Japan such investigations into the problem of coal have not appeared as in England or the United States, the examination of overseas trends has gradually produced debate on natural resources and their conservation, and from Meiji 43 [1910] to 44 [1911] the Bureau of Mines and the Mining Inspection Office, both under the Ministry of Agriculture and Commerce, conducted a survey of the amount of coal deposits in Japan. This was the first survey on matters concerning coal in our land, the general details of which appear in [the Ministry's] *Outline of the Coal Survey*.*

During the great war in Europe, starting with England and the United States, the various combatant countries boisterously debated matters relating to coal, petroleum, and other fuels, and this also led Japan to an ever-greater volume of discussion on the problems of coal and petroleum. Outside of the government there is Naitō, as well as the Fuel Research Institute's Tsujimoto and Yonekura,† the latter two strategizing the setting up of said institute and

---

*Most likely a reference to *Sekitan chōsa gaiyō*, ed. Nōshōmushō Kōzankyoku (Tokyo: Nōshōmusho Kōzankyoku, 1913 [Taisho 2]), https://dl.ndl.go.jp/info:ndljp/pid/942529.

†All engineers of repute in the late Meiji to early Shōwa period: Naito Yū (内藤游); Tsujimoto Kennosuke (辻元謙之助); Yonekura Kiyotsugu (米倉清族). See Seow, *Carbon Technocracy*.

other such organizations. They have each conducted research relating to coal and made no small contribution to the general fuel industry. And separately, the Ministry of the Navy has established a fuel depot at Tokuyama [in Yamaguchi Prefecture] that mainly manufactures briquettes with an affiliated research department that engages in various kinds of related projects.

In August of Taishō 9 [1920], the government installed the Fuel Research Institute in the town of Kawaguchi, Saitama Prefecture, with a new operating budget of 1,000,000 yen, placing it under the jurisdiction of the Ministry of Agriculture and Commerce with the mission of exclusively investigating and conducting research on fuel-related matters.

3   The Fear of Being "Left Behind in the Dust"

The Rise and Potential Fall of Coal in China

SHELLEN X. WU

CURRENTLY CHINA BURNS HALF the coal consumed on the entire planet and emits a third of the world's emissions. The dominance of coal is a legacy of China's entry into the modern world order. At the same time, while the use of this carbon-rich fuel is still rampant, China has signaled its full commitment to decarbonization: Five years after the 2015 Paris Climate Accords Agreement, President Xi Jinping announced to the United Nations General Assembly his party's intention to cut China's net carbon dioxide emission to nearly zero by 2060. To do so, China's coal consumption will have to peak sometime around 2030. While there remains uncertainty around whether and when this peak will happen, China has actively begun to revamp its energy infrastructure; the country has developed the largest total renewable capacity on Earth and, to a large extent, controls the global alternative energy supply chain. If China is to stay on its announced path, the long rise of coal power in the Chinese energy economy will have to end in the coming years. However, the growth of renewable energy capacity in China is occurring in step with the approval of new coal-powered infrastructure, resulting in parallel development rather than a transition from or replacement of the carbon-based economy. Coal was the accelerant that propelled China to become an advanced industrial economy, and yet today, despite ongoing growth in this fuel's consumption, the implications of the state's relation to coal have fundamentally changed.

How was China's dependency on coal achieved? In 1919, the National Geological Survey of China was formally launched, marking a step toward a reassessment of the nation's productive potential for coal-powered industrialization. The event was marked by the publication of the *Bulletin of the Geological Survey of China*. The inaugural issue contained multiple articles that surveyed coalfields. The selected translation presented here comes from this journal, which was established at a key turning point in the founding of both the scientific discipline of geology and the history of energy in China. The Qing empire had collapsed in 1911, bringing about an ignominious end to several thousand years of imperial history and leaving a power vacuum in its wake. For most of the 1910s and 1920s various political factions jockeyed to reunify an empire that had fractured into numerous scattered power bases and enact their vision for the future of China. In various popular publications of the time, writers regularly expressed fears that China would follow the fate of countries like Poland and the Ottoman Empire and disappear as an independent entity.[1] Western powers like Britain, France, the United States, and the Soviet Union seemed poised to take advantage of China's weaknesses to monopolize and exploit the country's natural resources. As mining engineer Zhang Yiou (張軼歐) stated in another of the *Bulletin*'s prefaces, the fear was that without sufficient exploitation of its geological riches, China would be "left behind in the dust."[2]

Against a backdrop of deep political uncertainty, the combination of political and scientific interests expressed in the preface of the inaugural *Bulletin of the Geological Survey of China* provides historical detail that helps to explain the unique importance of coal in China as it transformed from a useful everyday energy source into the essential fuel of industrialization, and as a result of this, it helps explain how China's potential geological abundance became a measure of the nation's potential place in the world. Why is coal such an important part of China's political and economic discourse? What role did the science of geology play in fostering a growing nationalism? At the time, answers to these questions could be parsed from the goals of the national geological survey. A preface from the inaugural issue of the *Bulletin* is presented, alongside a foreword from the same issue: two windows into past views on coal. China did not follow the paths of other resource-rich but colonized countries around the world; even when deeply fractured and vulnerable, the country retained control over its mineral resources, particularly its significant coal deposits. The so-called resource curse, the paradoxical phenomenon whereby resource-rich nations find their development stymied as their wealth is exploited by others, did not strike China, despite considerable Western interest in its coal reserves in the nineteenth century and oil in the twentieth century.[3] From our contemporary perspective however, as a corollary to its developmental role, the primacy of coal in China has had significant negative consequences for both the local and the global environment.

There is a long history of mining and interest in mineralogy for both aesthetic and practical purposes in China. Strange and beautiful rocks, purchased at considerable expense and sometimes transported over long distances, decorated the gardens of the wealthy. Moreover, people had long and extensively mined and burned coal for smelting and heating in the north, extracted and traded copper and silver for use as currency, and had pioneered the use of iron for farming tools. As noted in this volume's introductory chapter, a comparatively advanced and large-scale eleventh-century Chinese coal-fired metallurgy industry was rediscovered in Western economic history through the work of Robert Hartwell.[4] Hartwell's work fueled later debates about a "Great Divergence," the idea that non-Western regions had the capability to take off into self-sustained economic growth by the mid-eighteenth century but, for a variety of reasons, did not.[5] By the late nineteenth century, interest in the science of geology had been encouraged by the fact that the knowledge it had amassed was the key to unearthing valuable resources and that its principles could inform the practicalities of mining and mineral resource development. Intellectuals and political leaders alike viewed Chinese control over these resources as essential for the country's continued independence and argued that the state should take up the science of geology. These thinkers had reached this conclusion because of clear Western interest in China's mineral resources. From the 1890s, foreign consortiums had regularly dispatched teams of surveyors to the Chinese countryside to assess the coal and mineral potential of specific regions and to lobby provincial governments for exclusive mining contracts.[6]

The first issue of the *Bulletin* carried a foreword in both English and Chinese by the director of the survey, Ding Wenjiang (丁文江; 1887–1936), a

University of Glasgow–trained geologist who had been appointed director of the country's geological survey at the precocious age of twenty-six and later served as secretary-general of Academia Sinica, the Chinese academy of sciences, founded by a newly established national government in 1928.[7] Two additional prefaces, only in Chinese, by Minister of Agriculture and Commerce Tian Wenlie (田文烈; 1853–1924) and the mining official Zhang Yiou (張軼歐) were commemorative articles of a kind common in inaugural issues of such publications. The research articles in this first *Bulletin of the Geological Survey of China* surveyed coalfields in Anhui, Zhejiang, and Shanxi.[8]

In his English-language foreword, Ding mentioned previous geological surveying work done by the German geologist Ferdinand von Richthofen (1833–1905) in his travels in China from 1868 to 1872. Through a series of well-known publications, based on extensive fieldwork during those four years, Richthofen brought China into the folds of a select community of nations around the world that could claim a new and scientific authority in having surveyed and quantified their mineral resources.[9] His publications offered the first geological depiction of China in the West, and Richthofen subsequently became the leading European expert on Chinese geology. He believed the abundance of coal in China that his survey had uncovered would allow the country to industrialize under the tutelage of European powers.[10]

In this inaugural issue of the *Bulletin of the Geological Survey*, the director of the survey Ding Wenjiang opened his English-language foreword with a quote from volume one of Richthofen's works on China (see "Foreword" by Ding below). It was a derogatory comment about the unwillingness of Chinese literati to engage in strenuous physical activity.[11] Ding had perhaps imagined that Richthofen's words might spur the development of Chinese geology through the shaming of his compatriots—we can only speculate. In any case, by the early twentieth century, political leaders and student revolutionaries alike indeed viewed it as an embarrassment that a foreigner had so thoroughly studied China's resources while they themselves remained ignorant of their nature and extent. Chinese students studying abroad wrote to government ministries informing them of coming across Richthofen's surveys overseas. However, as it was, the research articles that were presented in the first *Bulletin* on coalfields and subsequent editions indeed mirrored Richthofen's work from decades earlier.

In the late nineteenth century, news of Richthofen's high praise for the extent and quality of Chinese coal quickly spread among Qing merchant-officials, even when they did not refer to him by name.¹² Richthofen's exploration of the relatively remote Kaiping area of northeast China (today Hebei province) played a role in the founding of China's first modern colliery, an infrastructural development replete with mechanical winches and pumps and artificial means of ventilation; the development also encouraged the construction of the first rail tracks in China in that area in the 1870s.¹³ Acknowledging the Europeans' developmental influence, the merchant and comprador Zheng Guanying (1842–1922) wrote in 1892:

> All countries rely upon their mineral products to attain wealth. England has the most plentiful mineral products, and therefore the country is also the wealthiest. Once a Westerner said, "Shanxi has coal deposits across 14,000 li, with approximately 73 hundred million megatons of coal. If all countries under the heavens use 300 megatons of coal per year, then Shanxi alone can supply the world for 2,433 years. Moreover, most of the coal is anthracite, and harder than American anthracite."¹⁴

The information clearly came from Richthofen's original letters to the Shanghai Chamber of Commerce, which had funded his expeditions in China, and the "Westerner" Zheng referred to in his writing was Richthofen. The widespread circulation of the German geologist's assessment of China's potential coal reserves from the late nineteenth century onward is also attested to by the fact that in the inaugural issue of the *Bulletin*, both the scientific leader of the geological survey and the political appointees of the Ministry of Agriculture, Industry and Commerce referred to him.

In the decade before the launch of China's first geological survey, a new generation of Chinese geologists had already begun agitating for the establishment of a rigorous stocktaking of the country's resources. This group of largely foreign-educated geologists later served as part of the survey team and counted some of the most prominent scientists of the era among their numbers, including Ding Wenjiang, Zhang Hongzhao, Xie Jiarong, Weng Wenhao, and Li Siguang.¹⁵ Xie's, Weng's, and Li's careers continued after the People's Republic of China (PRC) was established in 1949. Li Siguang is still celebrated today as one of the country's most prominent scientists. A few had also prepared for the civil service examinations, the traditional career path to officialdom for promising young men until it was abolished in 1905. Zhang Hongzhao, for example, passed the first level of examinations before going to Japan to study.

Ding Wenjiang regularly turned to classical Chinese records for inspiration. Although the civil examinations primarily focused on knowledge of such classics, there was a longer history of official interest in practical learning, grouped under the heading of the statecraft school. Late imperial officials subscribing to a statecraft school of governance emphasized mining, sericulture (silkworm cultivation), and agriculture as worthy ways to foster local prosperity.[16] The term *practical learning* (shixue 實學) comes from a Neo-Confucian concept that by the nineteenth century was adapted to include Western learning, including science and technology.[17] There was a clear imperial precedent for official interest in and sponsorship of knowledge with local and practical applications. A change occurred in the late nineteenth century, as instead of industries being affiliated with localities across the empire, coal mining and its related forms of resource extraction and fabrication came to represent a defining feature of an expansive modern nation. As I have argued elsewhere, geology played a key role not only in establishing China's new role in the global order but in also transforming its domestic politics.[18]

In a number of very important ways, Chinese scientists viewed themselves as continuing in the leadership roles of the literati in the imperial era. In the 1910s, the first group of Chinese scientists to receive training in geology agitated for state support for establishing a geological survey. In their petitions, they noted the various foreign geologists' assessment of China's vast coal potential and argued that an independent national survey was essential for maintaining the country's sovereignty and control over its mineral resources. From the beginning, scientific discourse on coal was entangled with a nascent nationalism and a recognition that the country's political sovereignty relied on maintaining control over the mineral resources that were essential for a certain type and desired scale of industrialization. China's first generation of professional geologists promoted their field as an essential form of knowledge for the protection of national sovereignty over its natural resources.

Under Ding's leadership of the geological community and the survey, Chinese geology became internationally recognized.[19] At a time when government funding for the sciences was uncertain and conditions at public universities and laboratories were often quite poor, geologists managed to build an infrastructure for the training of students in the field and undertook some of the major scientific surveys of the early twentieth century, including the discovery of the Peking Man fossil at Zhoukoudian in 1926, which kicked the study of paleontology in China into high gear.[20] The structure of the *Bulletin*, with an English table of contents and abstracts from the back and a Chinese table of contents and articles from the front (read in the vertical, right to left, Chinese style), reflected its author's intended inclusive approach and hoped-for international influence.

The English preface by Ding provided a concise overview of how, by 1916, the geological section in the Ministry of Industry had become an independent geological survey unit located in a separate building and with its own budget. However, interestingly, only the exclusively Chinese preface, reproduced here, written by the head of the Ministry of Agriculture and Commerce Tian Wenlie (1858–1924), explicitly connected geology with the need to develop and exploit China's extensive mineral wealth. Tian pointed out that geology was closely related to mining and argued that unless the nation began actively exploiting its resources, it would be as if they were "letting valuables lie about and inviting thieves in."[21] Politically, Tian was the most conservative of the three men who wrote these introductory essays. He was closely aligned with the first president of the Republic, Yuan Shikai, who, before his death in 1916, had attempted to restore the monarchy. Yet, despite these disparate political views, Ding, Tian, and Zhang agreed on the importance of mining for this new nation.

The backgrounds of these three men, each of whom contributed to the prefaces of the inaugural issue of the *Bulletin,* reflect the political spectrum of the era: from Tian, the conservative carryover from the imperial era, to Ding, a leading member of a new generation of foreign-educated scientific elites. These multiple political visions for China overlap in one area—a broad agreement on the importance of maintaining control over and developing the country's mineral resources, including, most importantly, its abundant supplies of coal. Tian repeatedly compared China's mineral deposits to foreign locations like Germany's Ruhr Valley, an area well-known for its abundant coal and coal-powered industries. His support for the science of geology rested on its importance for the development of the mining industry and ultimately as a means to attain wealth and power. The developmental potential of this still-hidden wealth would provide the way for China to climb to parity with Europe and the US. This singular focus on the use of natural resources informed the orientation of Chinese science and made geology at once a highly international but also highly nationalistic science.

From the nineteenth century onward, coal shaped the political and historical discourse of China. Yet our understanding of coal's role in twentieth century China has been shaped by successive narratives, which generally downplayed, if mentioning at all, the environmental impact of the country's dependence on coal for both industrial and domestic use. Surprisingly, despite the early communist organization of miners, for example, at Anyuan Coalmines in Jiangxi province, few sources make any mention of labor or environmental consequences.[22] Instead, as with earlier discussions from the late nineteenth century onward, political leaders, intellectuals, and geologists talked about ways to combat imperialism and foreign nations' efforts to control Chinese mining resources. Coal and other natural resources were presented as parts of a wider policy of anti-imperialism, which continued in the early decades of the twentieth century, the Republican period, and in the People's Republic of China after 1949. In the 1950s and 1960s, historians portrayed late Qing industries and mines as part of a feudal stage of development. While political leaders in the communist regime used their experience in organizing miners in the 1920s to affirm their revolutionary status, miners themselves, their health, and the environment they lived in remained largely unreported.

From the 1910s, the geological survey and studies of China's natural resources reduced coal to a numerical measure of progress. The quantitative measure of Chinese coal reserves signaled potential prosperity, and their potential consumption promised wealth and power. Yet those numbers foretold little about coal's place in China's later ecology, understood in both an industrial and environmental sense, and the connections that coal-derived productive power would create with the rest of the world.[23]

The assumed coupling between increased coal consumption in China (and elsewhere) and increased economic progress has, in more recent years, become highly contested. Air pollution and climate change impose clear limits to this and other fossil-fueled economies. Today, Chinese industries' dependence on coal is blamed for much of the marked effects of air pollution on human and environmental health.[24] The issue of pollution is now recognized as a serious problem and the Chinese government has voluntarily agreed to reduce its rate of coal use. However, far more drastic reductions will be necessary. A recent reduction in air pollution in Beijing came not from the reduction in coal-fired energy production in the country but from the removal of polluting factories from the vicinity of the capital.[25]

As stated in the introduction to this book, ahead of the 2015 Paris Conference, China pledged to reach peak emissions by 2030 by cutting greenhouse gas emissions and increasing the use of non-fossil energy—an agreement to which Xi Jinping added that China would reach "net-zero" by the year 2060 in his UN General Assembly speech in 2020. There is now concrete action in China to ramp up renewable energy supply and change its energy infrastructure, from homes to power plants and furnaces across the country.[26] China controls the supply chains of alternative energy technologies from batteries and wind turbines to photovoltaic modules. At the same time, the country's leaders continue to permit the construction of coal-fired power plants and encourage the spread of Chinese fossil energy industries overseas, in places such as Pakistan and Kenya.[27] When it comes to the question of protecting the environment or pursuing economic development, the latter has proven the priority. Like other large emitters, the government in Beijing has consistently promoted coal and other fossil fuel industries. Of course, China is not alone in favoring short-term economic growth over long-term environmental protection. The report of the sixth Intergovernmental Panel on Climate Change (IPCC) makes clear that measures undertaken by most countries are grossly insufficient to curb the most serious effects of climate change.[28] We have entered an era in which expanded coal use takes place alongside efforts to decarbonize. Geopolitical struggles over photovoltaic panels, wind power, and rare-earth minerals accompany those over coal.[29]

The history of coal in twentieth-century China encapsulates the problem at the heart of environmental history. Is environmental history the histories of distinctive ecologies or the cultural history of how humans interact with nature, itself a loaded and ambiguous term? What is the role of human agency in this history?[30] What it means to use coal has changed over time just as its significance to the Chinese state has shifted with the country's industrialization and now its push for alternative energy. Over the course of the twentieth century, the discourse on coal and other natural resources was successively made to fit into larger ongoing concerns. The Japanese invasion of China in the 1930s reinforced this propensity as the wartime state moved its energy concerns to border areas.[31] By the 1970s, with the expansion of capitalism, we encountered the exhaustion of natural resources and new frontiers, both on Earth's surface and in its subterranean realms. In the last decade, energy prices have swung sharply, depending on a confluence of geopolitical factors. Deep-sea drilling and fracking, as well as geopolitical realignment of new resource frontiers in Brazil and Russia, resulted in the virtual collapse of oil prices by late 2015. The global pandemic of 2020 exacerbated the collapse of oil prices, only for them to rebound afterward. Paradoxically, low energy prices failed to stimulate global economies and were instead portents of further global market turmoil and deepening recession in developing countries dependent on China's seemingly insatiable energy demands. In 2022, Russia's invasion of Ukraine and rigorously enforced COVID-19-related lockdowns in China suddenly and dramatically brought energy prices to new heights. In response to disruptions to supply, China once more turned to coal to sustain growth. As China faces the environmental consequences of industrialization, it replicates nineteenth-century European empires by exporting its environmentally damaging resource infrastructures to its peripheries and overseas. Today, China is both the largest total consumer of coal and the largest producer of renewable energy. Rather than being left in the dust, the challenge is to get rid of the dust of coal-powered development. Coal's journey, it turns out, is also the journey of modern China.

TIAN WENLIE (田文烈)
"Preface" (1919)*

Geology is closely related to mining. The recent development of the mining industry reflects the application of geology. Although the wealth of mineral production in our country is praised around the world as inexhaustible, the location of mineral veins and the extent of deposits have not been ex-

---

*Tian Wenlie (田文烈), "Preface," *Dizhi huibao (Bulletin of the Geological Survey of China)*, no. 1 (1919), 1.

"Preface" (1919)

amined. There was only the German geologist Ferdinand von Richthofen's *China* works, which broadcast abroad information about Chinese minerals. By contrast, we only heard about this material from abroad. Is this not letting valuables lie about and inviting thieves in? Moreover, should we not be faulted for not learning this discipline?

Since antiquity, China has practiced mining. The mining of gold and other metals dates to antiquity. According to the legends, since the time when the Yellow Emperor defeated Chiyou (the God of War) the Chinese have made metal weaponry. Stone vessels were replaced by various metallurgical creations in the Shang and Zhou. By the Qin and Han dynasties, metals were used for knife money (cast bronze, knife-shaped money). From our ancestors we still have unrivaled records of regulations. Those who mined accumulated knowledge from experience. However, many mining practices were still limited to the margins of knowledge.

Still today, miners work toe to heel in native mining pits with only enough room to open one's eyes. Because the Chinese mining industry relies on such fragile means [because of the lack of state support], the results can be observed. Today, everyone who thinks about opening enterprises, many times will hurry to the mining industry. Not necessarily waiting for government's incentives, those who look to operate mines may not be fortunate enough to find it worthwhile. While seeking for a location where to begin work, many make multiple false starts.

In seeking to elevate a country's people, many see the most effective way as having the mining industry leading the way. For such efforts to be successful, the responsibility lies with those who set the policy to first support geological survey and to investigate where the results will be satisfactory, and then to solicit mining efforts. Therefore, the Ministry of Agriculture and Commerce established a geological survey institute, which submits yearly reports with accurate details. Now we have gradually accumulated works with clear illustrations and explanations. Those who rely on this publication [*Bulletin of the China Geological Survey*] seek guidance.

Those engaged in mining should follow this research to guide their efforts; it will lead to results that are twice as effective with only half the effort. They should seek the path that will lead to the desired outcome. Meanwhile,

Tian Wenlie (田文烈)

investigators who toil all year round in trenches and endure the hardships of their labor will also see rewards, and their efforts will not be in vain. Those who produce inferior works are not necessarily at fault. Geologists rely on boats and other means of travel. But state funding frequently falls through without warning and given the vast distances in our country, we can hardly fault those who avoid going to remote places.

Of all the mineral deposits, coal, iron, and petroleum are the most in demand. Of the coal fields discovered so far, Kairong is the largest, but compared to the German Ruhr region is only a tenth of the size. The known iron deposits in the country measure a mere 500 megatons, 7 percent of the output as compared to the French mines of Alsace-Lorraine. Shaanxi province is seen as the best location for petroleum. This year the province was surveyed by the American Mobil company, which showed that the potential will not be of great value. As for silver, and copper mines, these have not been fully evaluated. These are the fruits of our so-called inexhaustible mining potential.

Yet our country's greatest source of wealth lies with mining. Fortunately, more is yet to be discovered. For those who look to develop our country and aim for economic independence and who seek the survival of our society must look to mining. For this to happen, we depend on researchers' investigation to fully exploit these resources. The countries that sponsor surveys benefit from this work. If we look to support the mining industry, we will one day be able to compete with Euro-American countries.

July 1919. Minister of Agriculture, Industry, and Commerce Tian Wenlie

DING WENJIANG (丁文江)
"Foreword" (1919)*

... da der chinesische Literat schwerfällig ist und für die schnelle Bewegung ein fortdauerndes Hinderniss bietet, ... und sich von den landesthümlichen Vorurtheilen über das Decorum nicht frei machen kann. Zu Fuss zu gehen ist in seinen Augen erniedrigend, und die Beschäftigung des Geologen ein directes Aufgeben aller Menschenwürde.†

RICHTHOFEN, *CHINA*, VOL. 1, P. 38

The quotation was written by Richthofen in 1877 and 35 years later geology first gained recognition in this country as a necessary branch of the Government services; for in 1912 there existed a Geological Section in the Bureau of Mines which formed a part of the Ministry of Industry in the Provisional Government at Nanking. It continued to exist as such when the seat of government was transferred to the north. In 1913 I was appointed chief of the section and was entrusted with the organization of a real Survey. Now as far back as 1910 the Government University of Peking started a department of geology and had Dr. F. Solger of Berlin as its principal professor. The Department was however not quite a success owing to the scarcity of students, and there was a proposal to discontinue it. With the help of Mr. E. Chang, then Director of the Bureau of Mines, I obtained permission from the University

---

*V. K. Ting [Ding Wenjiang] (丁文江), "Foreword," *Dizhi huibao (Bulletin of the Geological Survey of China)*, no. 1 (1919), unpaginated. Translated by Shellen X. Wu.

†"... because the Chinese intellectual is cumbersome and a constant obstacle to rapid movement, ... and cannot free himself from the national prejudices about decorum. In his eyes, walking is humiliating, and the occupation of geologist is a direct abandonment of all human dignity." Translated by the editors.

authorities to take over the whole department and establish in its stead a school of geology under the joint auspice of the Ministry of Agriculture and Commerce and the Peking University. Mr. H. T. Chang, who in fact first made the proposal of training our own staff, became director of the new school. Dr. F. Solger was reengaged to do part of the teaching and also as a member of the Geological Section, but he left in 1914 when the war broke out, and most of the teaching work fell upon Dr. W. H. Wong who came to Peking towards the end of 1914.

In the beginning of 1916 the old Geological Section in the Bureau of Mines was reorganized into an independent Geological Survey with a separate building and a separate budget. In the summer of the same year the students in the School of Geology graduated and 18 out of 30 were appointed as junior members. The University of Peking reopened its department of geology and the new Survey has therefore nothing more to do with education. Very little change has taken place since that date, and at present the Geological Survey of China has the following staff:--

<pre>
Director   V. K. Ting
Sectional Chiefs:--
                    H. T. Chang
                    W. H. Wong
Geologists:--
                    L. Wang
                    Y. T. Wang
                    L. F. Yih
                    T. C. Wang
                    H. C. Tan
                    T. O. Chu
                    C. C. Liu
                    M. H. Li
                    T. H. Chao
                    C. Y. Hsieh
                    T. H. Chow
                    W. M. Hsu
</pre>

"Foreword" (1919)

                    T. I. Loo
                    H. T. Li
                    C. Li
                    J. C. Chao
                    P. Y. Tung

Budget 80,000 dollars (Chinese currency).

In 1914 Dr. J. G. Andersson, then Director of the Swedish Geological Survey, was appointed Mining Adviser to the Chinese Government, and his two assistants Messrs. F. R. Tegengren and E. T. Nyström became members of the Geological Survey in 1916, but they both left the next year, since which time there has been no foreigner on the staff. But Dr. J. G. Andersson has been intimately connected with the work of the Survey though technically he is not a member.

It has been planned to publish 3 series of maps:--
(1). Maps covering the whole of China on the Scale of 1 : 1,000,000.
(2). Maps on the scale of 1 : 200,000 along the navigable rivers and railways.
(3). Detailed maps of special regions either important for mining or otherwise scientifically interesting

The accompanying publications shall be of 2 series (1) bulletins containing smaller papers and miscellaneous material, (2) memoirs in which the more serious studies shall be included. Every paper published in Chinese shall either be translated into one of the principal European languages or at least a summary in that language shall be given in order to make it accessible to the general scientific public.
We have now in the press:--

Memoir series A. No. 1. Mineral Resources of China with a geological map of 1 : 8,000,000.
Memoir series A. No. 2. The Geology of the Western Hills of Peking.

Ding Wenjiang (丁文江)

Memoir series B. No. 1. Geology and Mineralogy as known to the Chinese.

It is hoped that bulletins No. 2 and No. 3 as well as the following memoirs will be ready next year:--

1: The Geology of Shantung with a general map of 1 : 1,000,000.
2: The Geology of Shansi with a general map of 1 : 1,000,000.
3: The Iron Resources of China.

Any help or criticism from our colleagues abroad will be greatly appreciated.
    Peking, December 1919. V. K. Ting [Wenjiang's English pen name]

4 Frederick Tryon and the Decoupling of Energy and Economic Growth in the 1920s

ANTOINE MISSEMER

SINCE THE 1940s, many countries have measured economic growth according to Gross National Product (GNP) or Gross Domestic Product (GDP); accordingly, economic growth, measured in terms of the value created by the production of goods and services, has been one of the main objectives of economic policy.[1] Since then, evident environmental problems and the accelerating impacts of climate change have challenged the logics of the continued pursuit of economic growth. However, policymakers and citizens have not significantly altered their commitment to this kind of economic growth. In part, this is because a hope—or belief—has emerged: that it is possible to *decouple* economic growth from energy consumption and the carbon dioxide emissions associated with combustion. Decoupling suggests humankind can achieve a sustainable kind of growth alongside continued economic expansion. Decoupling, it is argued, has already begun and will continue to occur, in part thanks to policy measures but also due to the market's ability to incentivize the development of less energy-intensive consumer behaviors, technologies, products, and services. Given the continued pursuit of economic growth, the veracity of decoupling, as a means to achieve a sustainable kind of development, is fundamental to humanity's future; as such, it has become a topic for debate in a range of disciplines.[2]

Perhaps surprisingly, the roots of the decoupling debate do not begin in the 1970s, during a period riven by energy crises, as one might expect, nor do they begin in the 1980s, with the rise of the notion of "sustainable development."[3] In fact, seemingly the earliest instantiation of the principles of decoupling can be found in the 1920s, when the North American economist Frederick G. Tryon published his article "An Index of Consumption of Fuels and Water Power" in the *Journal of the American Statistical Association*. Tryon's goal was to measure the relationship between energy consumption and economic output. In establishing this correlated relationship, he noticed periods in which energy and productivity appeared to be separable functions. This encouraged him to speculate on the possibility that the relationship between energy and economic growth might have a significant degree of flexibility. However, beyond his immediate readership, Tryon's contribution and this specific observation went largely unnoticed until the 1970s. In recent decades, the paper has been mentioned in passing, without detailed examination.[4] With the exception of economist Ernst Berndt in 1978, Tryon's work has only recently been subject to more thorough analysis.[5]

In the United States, proponents of the Progressive Era (1901–1909) conservation movement, a national political agenda that sought to encourage the fledgling nation to use its fecund resources with greater care, raised concerns among the public and the political class about the finiteness of natural resources and the need to safeguard the nation's remaining wilderness.[6] Forester Gifford Pinchot, arguably the movement's intellectual leader and certainly its most vocal advocate, strongly encouraged economists to get involved in issues of resource and wilderness conservation. The result was that from the 1900s to the 1920s, a number of what can be seen as proto-environmental economists emerged: Bernhard E. Fernow, Lewis C. Gray, Richard T. Ely, and John Ise. They established a range of conservationist economic principles that would be recognizable to today's economists and policymakers, such as optimal rates of resource extraction, intergenerational equity, and fiscal instruments for resource substitution. However, their contributions did not have a decisive influence on mainstream economic theory, which became more concerned with the pressing issues of maintaining economic production, controlling inflation, and sustaining employment.[7]

The mechanized forms of combat and oil-intensive technologies used in the First World War had clearly played a role in asserting the centrality of energy to the maintenance of state power. The energy demanded by these technologies also encouraged the state to intervene in specific sectors of fuel provision.[8] Perhaps as a result of this, at the turn of the 1920s, energy became a specific concern of economists in the US, particularly insofar as it lent itself to sectoral studies of the economy that examined a single resource or product. Economists Joseph E. Pogue and George W. Stocking wrote on petroleum, for instance.[9] John Ise devoted a volume to the economics of forestry and a further volume to the economics of oil.[10] On a more theoretical ground, the mathematical economist Harold Hotelling attempted to develop a comprehensive approach to the economics of exhaustible resources, focusing primarily on petroleum.[11] Despite such sectoral-focused contributions, the field of energy economics, understood as the study of the overarching productive and profit-making relationship between different fuels and power sources, was not yet an established discipline.

Economists were not alone in tackling energy issues in relation to production activities. Engineers also worked on the role of power, understood as the applied use of energy, particularly in the form of electricity, as a driver of economic growth. It is from that perspective that the 1924 World Power Conference, the ancestor of today's World Energy Council, was established. The aim of this organization was to take a global view of the energy economy, primarily from the perspective of engineering.[12] This trend would resonate with the technocrat movement of the 1930s, a US-based populist movement, in which economists and engineers began to think about the energy values of goods and services.[13] In contrast to economists, engineers generally did not confine themselves to sectoral studies; they dealt with energy more holistically, considering power as a whole and at varied scales.[14]

Tryon put forward a conception of energy that was quite close to that of engineers.[15] In his 1927 article, his analysis concerned power insofar as it addresses the interrelation of various fuels *and* water power. Throughout his text, his terminology testified to the centrality of energy in general to economic concerns: the word "energy" appears 54 times, "power" 46 times, "coal" only 15 times, "oil" and "petroleum" 5 times, "waterpower" 5 times, "gas" 3 times, and "firewood" twice. In both word and deed, Tryon was clearly an early energy economist.

Alongside these concerns, a more central reason why Tryon deserves to be included in the pantheon of *energy* economics is that this novel paper undertook a wide-ranging "reconnaissance in the field [of the economics] of power as a factor of production."[16] In doing so, the paper addressed a fundamental and as yet unresolved question, one that is still being puzzled over today and remains of great relevance to the ongoing climate crisis—namely, the fixity of the relation between energy consumption and economic growth.

Tryon (1892–1940) was a mineral geologist, an economist, and a statistician who primarily worked at the US Geological Survey and then at the US Bureau of Mines, from the 1910s to the 1930s.[17] Throughout his career he published dozens of articles and reports on mineral resources and on issues raised by population growth and industrial production.

In 1922, Tryon made contact with Robert S. Brookings, a businessperson and philanthropist who had become rich through selling homewares. Brookings was thinking about merging his three public-expertise research organizations—the Institute for Government Research, the Institute of Economics (which benefited from Carnegie Corporation support, the legacy of Andrew Carnegie's steel fortune), and the Brookings Graduate School of Economics and Government—into a single entity: the Brookings Institution. The organization would become the first think tank in the United States. Well funded, this Washington-based philanthropic research organization was intended to conduct policy-relevant research in all fields of public administration and economic affairs.[18]

Tryon became a part-time fellow at the Brookings Institution in the mid-1920s. Soon after, the institution launched a research program on what it termed "the power revolution."[19] Energy sources, in particular coal and waterpower, were already topics of concern at Brookings, with early publications supported by the institution.[20] Tryon was tasked with stepping up these efforts and bringing together economists and engineers to develop expertise on energy issues. Archival materials suggest that Tryon began working on this project in late 1926, aided by the recruitment of a statistician from the Bureau of Mines, Raymond Kenny, who would "compile certain figures of fuel consumption from various sources."[21] Their collective research endeavors resulted in a series of publications.[22] However, this work gradually lost some vitality in the mid-1930s, after the Brookings Institution ceased to support this energy-centered work because of other more immediately pressing research needs such as the unemployment situation during the Great Depression.

Tryon's 1927 paper occupied a special place in the Brookings energy program. On the one hand, it was explicitly part of the wider power revolution project, as Tryon points out at the opening of his article.²³ On the other hand, it seems quite clear that Tryon had only recently arrived at the organization when he wrote the paper, and so some of his ideas predated his collaboration with the Brookings Institution. In fact, Tryon had begun to be interested in amassing information about all sources of energy in relation to economic development after reading Chester G. Gilbert and Joseph E. Pogue's *America's Power Resources*.²⁴ His knowledge had also been furthered thanks to his attendance at a research seminar at the University of Pennsylvania in 1924–1925. There, Tryon had met Carroll R. Daugherty, a student, and invited him to reflect upon the idea of using horsepower-hours as a yardstick for all forms of driving force.²⁵ What was this unit? When British engineer James Watt had first invented his rotary steam engine in the 1770s, in order to help explain the utility of his product to potential customers, he had compared its power to that of the horses used to power mills. In doing so, he established a unit of power that remains a comparator today.²⁶

When Tryon arrived at Brookings, replete with his US Geological Survey background and his early inspirations about the role of energy in economic development, he met researchers involved in the emerging American institutionalist movement, including Walton H. Hamilton, Harold G. Moulton, Isador Lubin, and Edwin G. Nourse. Those scholars promoted a permanent back and forth between theory and empirical facts and insisted on public regulation of the economy.²⁷ This perspective was particularly well suited to both the theoretical and empirical ambitions of Tryon. At the same time, Tryon had developed his own trajectory in energy and economic affairs, and as such it cannot be said with certainty that he would have identified as an institutionalist. He shared several of the institutionalists' concerns, particularly with respect their enthusiasm for public regulation, but he showed no interest in another major institutionalist topic, primarily the related idea that there were limits to the rationality of economic actors and that the state should therefore guide their choices.

To better understand the relevance of Tryon's 1927 article, a close examination of his conception of the energy-growth nexus is necessary. In his analysis, as the title indicated, Tryon sought to build an "index of consumption of fuels and water power" that would allow for comparison between various sources of energy and that, in total, would allow for a more general index of the relation between power and economic output. The words *cou-*

pling or *decoupling* were not used by Tryon in this article; he instead chose to talk of "parallelism of the fluctuations" and the "close correspondence" between power consumption and economic growth.[28] What was at stake, nonetheless, was something like the modern conception of decoupling, not least because one of his graphs detailed the upward logarithmic curve of energy consumption and a similar curve that indicated the growing volume of production in the United States as a whole for the period 1870–1926.

The bulk of Tryon's article consists of a methodological explanation of the basis and construction of his energy index—that is to say, the presentation of his synthesis and measurement of energy consumption. The challenges were considerable, because the statistical series were incomplete and because there was no established consensus about how to aggregate different forms of energy (heat and power, for instance) together—the difficulties involved in finding reliable data were also commonplace in the engineering literature of the period. By choosing to opt for a common measurement in British thermal units (BTU), Tryon managed to provide estimates, which readers could find in the appendix of his article.[29] Regarding the energy-output index, Tryon stated that the correlation between the two series of data was "fairly clear"; energy consumption could even be seen, at times, to grow faster than economic output.[30] The First World War, he noted, seemed to have produced a "pronounced flattening of the trend," an anomaly that Tryon explained as a result of the fuel efficiency gains that were achieved during war because of fuel conservation policies.[31] However, rather than taking these initiatives as indicators of a coming age of conservation, Tryon argued the potential for future efficiency gains was low and affirmed his belief that economic production would remain strongly correlated to energy consumption in the future.

Tryon's statistical results appear quite robust for the period.[32] However, his ambitions were not only empirical but also theoretical.[33] If energy was considered as a factor of production, it was assumed to be a *cause* of production. This causal issue, or that of energy determinism, has been the subject of a multitude of works since the 1970s, and it remains largely unanswered, despite the sophistication of more recent econometric models.[34] Quite how energy and economic growth are correlated remains unclear. In his 1927 analysis, when he compared his energy series to other indicators, Tryon struggled to establish clear causal relationships. He has to admit, for instance, that the increase of transportation stimulated energy consumption, which would indicate that energy exploitation was not the cause but the result of more transportation options.³⁵ In reality, there are clear self-reinforcing loops between energy and output; their relationship is intertwined. The difficulties we still have today in fully understanding the energy-growth nexus were already present in the 1920s, which suggests that the question of energy determinism or indeterminism might be inextricable: energy flows and economic activities are too intertwined to be disentangled.

Tryon's overriding interest in this paper was the nature of energy supply and demand and the role of different fuels and sources of power in powering economic development. It seems, however, that his 1927 article also had other motives with respect to the impact of energy on labor. In the course of his argument, Tryon explained that he wanted to estimate "the total consumption of power in all forms, and the aggregate degree of replacement of human labor by power machines."³⁶ It is likely that the Brookings Institution had supported Tryon's project due to the light it might shed on the question of unemployment, which was seen to be rising dramatically because of the growing use of mechanical power in production.

A common if highly problematic metaphor at the time (and still in use today) was the description of power, especially fossil fuels, as something equivalent to "energy slaves," which akin to enslaved people could realize considerable amounts of work. One of the first occurrences of the term *slave* in relation to fossil fuels can be found in French economist Émile Levasseur's work from the late nineteenth century.³⁷ In the United States there was a long tradition of drawing comparisons between slavery and fossil fuel use that continued into the first decades of the twentieth century.³⁸ Tryon was aware of this metaphor, and it may have fed into his argument about the possible substitution of (paid or unpaid) human labor by mechanical power.³⁹

The idea that energy technologies can be detrimental to the goal of full employment is a problem directly related to the general relationship between technical progress and human labor. In that sense, Tryon's work can also be analyzed from the perspective of the long history of the social implications of technological change, between the loom-smashing Luddites of nineteenth century Britain to today's "tech" prophets who warn us that artificial intelligence may sooner or later come to replace human intellectual labor. Tryon does not seem to take sides in the defense or in the criticism of energy-labor substitution. He mentions only that measuring the extent of the phenomenon is of the utmost importance.

Throughout his career Tryon demonstrated attention to social issues and to the effects of technological progress on employment within the energy industries. Before his arrival at the Brookings Institution, he had already devoted a full study to the situation of mining workers.[40] In one of his last contributions, in the context of the Great Depression, Tryon had expressed concern for the consequences of widespread mine closures, partly because ore depletion, and the economic dearth it created, which risked leaving "stranded populations" of former mine workers throughout the country.[41] Regarding the specific question of the replacement of labor by mechanical power, he continued to be engaged in the subject until the very end of his career and was still writing on the topic in the late 1930s on the "social effects of pending technologic change" in the energy industries.[42]

By mixing an engineering conception of energy with economists' research questions and in initiating the reflection on the energy-growth nexus as an adjunct to the social consequences of energy-enabled mechanization of production, Tryon was somewhat of a disciplinary pioneer, addressing some as yet unanswered questions in energy economics. Let us not forget that he dealt with all these issues at once, in a single twelve-page article, of which extracts are presented here. Almost a century later, Tryon's contribution has therefore lost none of its relevance, with decoupling still occupying the center stage in contemporary debates about climate change and energy transitions, particularly as the social issues related to this transition are a central part of the many challenges to be addressed in the twenty-first century.

F. G. TRYON
"An Index of Consumption of Fuels and Water Power" (1927)*

Anything as important in industrial life as power deserves more attention than it has yet received from economists. The industrial position of a nation may be gauged by its use of power. The great advance in material standards of life in the last century was made possible by an enormous increase in the consumption of energy, and the prospect of repeating the achievement in the next century turns perhaps more than on anything else on making energy cheaper and more abundant. A theory of production that will really explain how wealth is produced must analyze the contribution of this element of energy.

These considerations have prompted the Institute of Economics to undertake a reconnaissance in the field of power as a factor of production. One of the first problems uncovered has been the need of a long-time index of power, comparable with the indices of employment, of the volume of production and trade, and of monetary phenomena, that will trace the growth of the factor of power in our national development. The problem presents many difficulties, and we are not sure that it can be solved, but a partial answer is given by constructing an index of the amount of the raw stuff of energy consumed, the coal, oil, natural gas, water power, and energy from minor sources that have been absorbed by the country.

[...]

---

*Frederick G. Tryon, "An Index of Consumption of Fuels and Water Power," *Journal of the American Statistical Association* 22, no. 159 (1927): 271–82 (extracts).

F. G. Tryon

REQUIREMENTS OF AN IDEAL INDEX OF POWER AND HEAT

[. . .] As a long-time measure of power as a factor in our national life, electricity is not satisfactory because it does not go back far enough,* and because it still covers only a part of the total amount of power generated. The extremely rapid increase in the output of the electric utilities—which averages about 10 per cent a year—does not give the net increase in the use of power because it includes the replacement of direct steam power. It does not even give the net progress of electrification, because many industries which formerly generated their own electricity are now purchasing current from central stations. Moreover, in spite of its rapid development, electricity is still only one among many forms of mechanical power. It has become the dominant form in manufacturing, and perhaps in mining, but it still supplies a very small part of the power used in construction, in transportation (except for the street railways), and in agriculture. [. . .] The significant thing for our purpose is not merely "superpower" or "electrification," important as they are, but rather the total consumption of power in all forms, and the aggregate degree of replacement of human labor by power machines.

Ideally, I suppose, what we want for an index of mechanical power is the total number of horsepower-hours generated by all forms of power equipment, no matter where installed or by what prime movers they may be driven. The attainment of such an ideal index is, of course, a long way in the future, but in the meantime something can be done by measuring the horsepower of the installed equipment. C. R. Daugherty, of the University of Pennsylvania, has just finished a notable paper which traces the total installed horsepower in the United States in all branches of activity, agriculture and transportation as well as mining and manufactures, at each census year from the Civil War to the present. Knowing the horsepower equipment and something of the average use factor and the amount of fuel consumed, it should be possible to make some estimate of the total number of horsepower-hours developed

---

*Electricity production by public utilities was reported for 1902, 1907, 1912, and 1917 by quinquennial censuses of electrical industries, but the continuous annual series does not begin until 1919.

"An Index of Consumption of Fuels and Water Power" (1927)

in the country that will at least show the trend for the last few decades.

Another approach to the problem of an index of power is to measure the amount of fuel consumed, or the fuel equivalent of other sources of energy. This is done in the series here presented. It will be said at once that such an index of raw energy consumed takes no account of improvements in the efficiency of converting fuel into mechanical power. That is very true, and it will shortly be pointed out how rapid improvements in combustion have affected the curve at certain points. But it is to be noted that improvements in fuel efficiency have been going on ever since Watt first took hold of the steam engine, and that the effects of a given improvement are spread over a period of years, so that the change from one year to the next is small. Its effect in the aggregate consumption of all energy is to alter the rate of growth, not to cause an absolute reduction in the amount of raw energy consumed.

[. . .] Heat and motion are twin forms of the same thing—energy—and either may be converted into the other. [. . .] The Industrial Revolution was no less a period of sudden advance in the art of applying heat than in the art of applying mechanical motion. We must think of the age of power, in a larger sense, as the age of energy.

An ideal index of power as a factor in production should therefore include heat as well as mechanical power. The two are interchangeable, derived from the same sources, and to an increasing degree supplied by the same agencies. Yet they are sufficiently distinct so that the ideal index should consist of two series, one representing direct heat, the other mechanical motion, the two being finally combined into a composite index of energy. The present index is a step, but only a preliminary step, in the solution of this problem. It includes the raw energy consumed for all purposes, heat and light as well as mechanical power.

[. . .]

COURSE OF THE INDEX, 1870–1926

In Chart I the index, uncorrected for annual growth, is plotted on logarithmic scale alongside an unweighted index of production worked out by Dr. Carl Snyder. [. . .] The parallelism of the fluctuations in the two curves is fairly clear, even in the period before 1890 [. . .].

F. G. Tryon

[. . .] Whereas the physical volume of production has been found to increase at the rate of something like 4 per cent a year, the consumption of energy over much of the period shown was compounding at the rate of from 5 to 7 per cent a year.

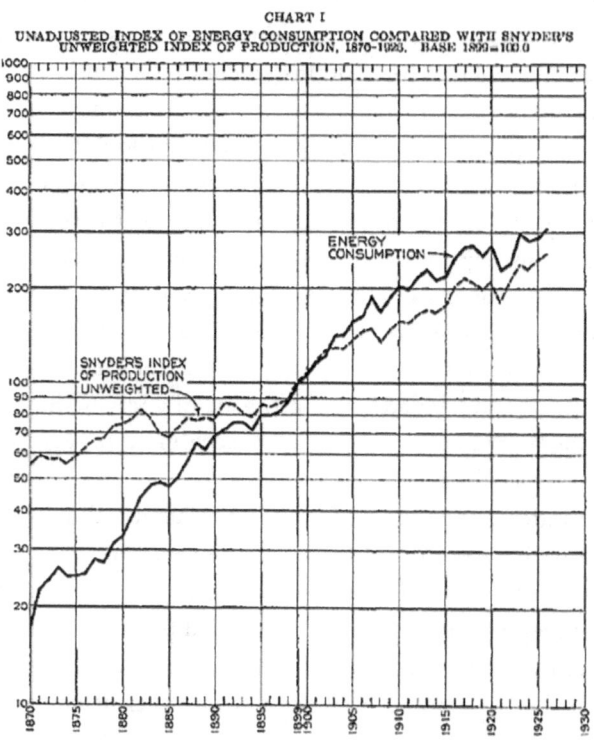

CHART I
UNADJUSTED INDEX OF ENERGY CONSUMPTION COMPARED WITH SNYDER'S UNWEIGHTED INDEX OF PRODUCTION, 1870-1926. BASE 1899=100.0

[. . .] It will [. . .] be seen that the rate of growth in energy consumption has altered its course during the period, as indicated by the changing slope of the curve. These changes are probably more apparent than real. The particularly steep slope in the period from 1870 to 1890 is partly due to the fact that the index does not include the energy of firewood, a relatively important source of heat and even power in the early days. At present firewood contributes only 6 per cent as much energy as the other materials, and the data on its use in the early years are too fragmentary to permit its inclusion in the index,

"An Index of Consumption of Fuels and Water Power" (1927)

except for the charcoal used in smelting iron. Enough is known, however, to suggest that were it included, the lower end of the curve would be raised and the early slope made less steep. The decline in relative importance of firewood was probably most rapid between 1870 and 1890. Again, beginning with the World War, a pronounced flattening of the trend is apparent. The high prices of fuel which began in 1916 and the actual shortages of the war itself stimulated interest in fuel economy and greatly accelerated the tendency to get more work out of the same quantity of coal, which had been present, though in a less degree, from the beginning. The change may be dated from 1917 [...]. It is not expected that progress in fuel economy can continue indefinitely at its present rapid rate, and in time the growth of energy consumption may be expected to resume a course more nearly parallel with that before the war.

RELATIVE GROWTH OF ENERGY AND OTHER INDICES

It is interesting to compare the growth of energy consumption with the growth of other measures of economic activity. [...] The increase in energy consumption was thus four times as great as the increase in population, and nearly twice as great as the increase in the total volume of production. [...]

The broad relationships are unmistakable. A great increase in per capita production is made possible by a still greater increase in power, and along with the process goes an increase in transportation which is the greatest of all consumers of power.

DEVIATIONS FROM THE TREND

Can this index of energy be used as a measure of economic activity? It appears to have small value as a forecaster, but when corrected for the growth trend it ought to be a faithful measure of the extent of a boom or a depression. [...]

5 The Colony and the World Energy Revolution

Meghnad Saha's Energetic Developmentalism

ELIZABETH CHATTERJEE

WHAT WAS NEW ABOUT industrial modernity? A pioneering generation of historians found an obvious answer in the realm of energy: the buried sunshine of coal and, later, oil allowed humans access to unprecedently vast quantities of useful energy, finally allowing societies to break free of the annual cycles of plant growth that had constrained premodern economic growth.[1] In recent years, though, many energy historians have dismissed the notion that there was any such sweeping transition—let alone the idea that such a transformation is a universal one through which all societies will pass. They have shown that the arrival of new energy sources has typically been slow and contested and that new fuels like coal did not replace "old" prime movers like firewood but rather joined them in an overall process of intensifying energy use; to cite one classic example, the importance (and size) of horses actually grew alongside steam power in nineteenth-century North American cities.[2] Postcolonial critics similarly reject the notion that societies across the globe are converging upon a single mode of energy-intensive modernity pioneered by the affluent West.[3] Looking more closely at the energy histories of different societies, these historians suggest, reveals at least as many continuities as changes and more diversity than convergence.

Yet, as they looked around the world in the first half of the twentieth century, many historical actors were convinced that an energy revolution had occurred—at least in some places. Even the avowed opponents of empire often believed that this world energy revolution was indeed a universal and linear path that all countries ought to follow, seeking a share of energy-intensive modernity for their own nations. Just as anticolonial nationalists staked a sincere claim to their own versions of modernity—without ironic scare quotes—so too did they develop their own distinctive accounts of the world-changing nature of new energy modes such as electricity. In the case of Meghnad Saha's "India's Need for Power Development" (1944), this accounting was literal.[4] One of the foremost scientists and public intellectuals of late-colonial and early postcolonial India, Saha developed a quantitative "energy-index," a metric for measuring the civilizational progress or backwardness of all countries within a single hierarchy. This was a diagnostic instrument deeply informed by his reading of the contemporary moment. Saha believed that there was an energy revolution that divided "medieval" from modern and that it explained much of the divergence between the "advanced" colonizing West and the languishing rest.

Saha did not deny the tight relationship between energy and industrial modernity established by the imperial West, then, though he preferred the hydroelectricity-rich model of a Sweden or a Canada to the less impressive fossil-fueled path of Britain. Instead, he sought to accelerate his nation's movement towards a singular energy-intensive modernity denied to the colony. He drew up plans to harness newly abundant energy in the service of a vision of state-led postcolonial liberation and large-scale industrialization. In his faith in the emancipatory potential of modern energy technologies as promoted by an interventionist national state, Saha represented a much broader current of electric developmentalist thought around the world. It is easy to cast him as the archetypal postcolonial modernizer, an uncritical champion of energy determinism and Western science in which "industrialism itself represented morality and culture."[5] Yet his writings on energy also offer up a series of more intriguing interventions in distinctively Indian debates around energy governance, the country's energy portfolio, caste and labor—not least in his deployment of the eye-catching metaphor of energy "slaves."[6] One of few lower-caste intellectuals in a still deeply elitist nationalist movement, Saha's commitment to social justice through abundant energy production and usage was inflected by a subaltern critique of opponents who romanticized the traditional organic energy regime, with all its sweat and inequalities. By his untimely death in postcolonial India, though, the impossible utopianism of his faith in state planning and energy modernization was becoming clear.

FROM THE STARS TO STATE PLANNING

Saha was born in 1893, the fifth son of a poor lower-caste shopkeeper in the village of Seoratali outside Dacca, in what was then the Bengal Presidency of British India (today Dhaka District, Bangladesh). He was a natural social critic. Legend has it that as a boy he changed his name to Meghnad after the son of the demon king Ravana in the *Ramayana*, a villain then newly reinvented as a hero by the poet Michael Madhusudan Dutt in a scandalous inversion of the Hindu pantheon.[7] If true, it was an apt transformation: Dutt's 1861 epic *Meghnad Badh Kavya* (The Slaying of Meghnad) both showcased a dazzling modernist fusion of Indian and European influences and, more ominously, the unjust fate of its loyal protagonist. A brilliant student, Saha attended Calcutta's prestigious Presidency College alongside a cohort of India's future scientific leaders, the first generation indigenously trained in the modern sciences. There he became involved with nationalist politics, even while contending with the caste discrimination that would plague his career. Saha would always be perceived by his more urbane contemporaries as something of a rough-mannered rustic, tensions worsened by his own vitriolic pen and "ambivalent relationship" with authority.[8] Barred from the imperial civil service examinations due to his associations with militant revolutionaries (though his actual activities are disputed), he turned instead to a life in science.

Physics was then enjoying an international golden age. Having taught himself German, the discipline's international language (in 1919 he would coauthor the first-ever book-length translation into English of Einstein's classic papers on relativity), Saha went on to make major contributions that, in the words of one historian "helped to change the course of modern astrophysics."[9] His elegant theory of thermal ionization, which drew on thermodynamics and the new quantum theory to link the structure of atoms to stellar spectra, was published in 1920. It gave him an international reputation, eventually securing his election as a Fellow of the Royal Society of London at the tender age of 34 despite anxious intelligence briefings suggesting that he was a "rabid revolutionary."[10] Though Western acceptance initially remained slow in the absence of experimental and observational data to confirm his findings, he would nonetheless be nominated seven times for the Nobel Prize in Physics in the decades that followed. After prestigious postdoctoral stints in London and Berlin—where he did indeed become a conduit for Indian revolutionaries[11]—in 1921 Saha returned to India to a full professorship at the University of Calcutta and then, fatefully, at the less well-resourced University of Allahabad. His scholarship thwarted by caste-inflected politicking and the persistent inadequacy of funding and experimental facilities, as well as his own tendency to pick fights with powerful rivals like the haughty Tamil Brahmin C. V. Raman, it would be as a public voice on science, an institution builder, and a social reformer that Saha would make his greatest impact. "I had lived in the ivory tower up to 1930," he later wrote, but "gradually glided into politics" to be of some use to the country in the all-important domain of applying science and technology to the greatest problems of national life.[12]

By the late 1930s, just as the Indian National Congress enjoyed its first brush with state power in the provincial ministries ushered in by a strained imperial regime, Saha was perhaps the most influential public scientist in India. He played a crucial role in the establishment of the Indian Science News Association and its popular periodical *Science and Culture*, created in 1935 as a forum for scientific and technical discussion with a broad audience. Over the next two decades Saha's 136 editorials covered a vast range of subjects from archaeology to calendar reform, while consistently advocating for close links between science and industry. He also helped to found the National Institute of Sciences (today the Indian National Science Academy). Devoted to river management and the generation of cheap electrical power, his presidential address to the Institute in 1938 concluded with an oft-quoted call for both ecological and social modernization:

> If we desire to fight successfully the scourge of poverty and want from which 90 percent of our countrymen are suffering, and lay the foundation of a strong and progressive national life, we must make the fullest use of the power which a knowledge of Nature has given us. We must rebuild our economic system by utilizing the resources of our land, harnessing the energy of our rivers, prospecting for the riches hidden under the bowels of the earth, reclaiming deserts and swamps, conquering the barriers of distance and above all, we must mould anew the nature of man in both individual and

social aspects, so that a richer, more harmonious and happier race may live in this great and ancient land of ours."[13]

In line with these preoccupations, Saha successfully lobbied Congress leaders for the creation of a committee to plan for India's reconstruction. Though the planning efforts languished when the colonial regime imprisoned many nationalist leaders during the war, by 1949 the various subcommittees had produced twenty-seven volumes of detailed prescriptions. The National Planning Committee established Saha as one of the country's most vocal and informed advocates of state-led developmentalism. Postcolonial freedom was only the means to a far more sweeping industrial and social transformation, not the end itself.

"Energy Slaves" and Indian Backwardness
Published in *Science and Culture* from a public lecture delivered to the Calcutta Rotary Club, which regularly hosted talks by leading scientists, Saha's 1944 piece "India's Need for Power Development" came out of his central role in the nationalist movement's planning efforts on power, fuel, and river management. It combined an energetic critique of British imperialism with his own deep commitment to science and social justice. At its heart lay the concept of what he called the energy-index, an aggregate measure of the population's annual "energy or work" calculated on a per capita basis that permitted systematic international comparison. Such comparisons were common at the time, appearing in the grim tables of energy use that littered the government's own plans as well as in books aimed at patriotic Indian children.[14] Interestingly, as the preceding chapter demonstrated, this kind of comparative strategy was mirrored elsewhere in the world by experts who sought to measure the extent of their own countries' energetic backwardness.[15]

Informing Saha's analysis was a distinctly linear, modernist sense of historical time and anticipation. In the words of *Science and Culture*'s first issue: "If the present is the child of the past, it may with equal emphasis be said that the future will be the child of the present."[16] The thread running through this civilizational history was the availability of energy. The "profound revolution" in vastly increased prosperity and material happiness in Western countries was explained by their rapid movement into what was popularly known as the "age of steam or electricity." Colonialism, too, was technologically determined. Elsewhere Saha looked back to the move from the Bronze Age to the Iron Age and to the fate of the New World's indigenous peoples after the arrival of the conquistadors: "The grand moral lesson of history is that, if a human community fails to take advantage of the newest technique for industrial production, it has no chance of maintaining its independence or individuality in the struggle with communities armed with superior technique."[17] The energy-index placed India and China at the bottom of the global hierarchy, their people still living in the "medieval" age of human and animal muscle power. Historical experience warned that unless a community believed in the idea of progress, it would suffer stagnation and subjugation—eventually vanishing entirely. It was a grim view of "machines as the measure of men" echoed by others within the *Science and Culture* community, testifying to the surprisingly pervasive impact of evolutionary sociology on radical Indian nationalism.[18]

This unpitying narrative of civilizational rise and decline informed Saha's equally unsettling energetic metaphor: slavery. A contemporary European or an American could boast "literally 10–15 slaves working constantly for him" in the form of steam- and electrically powered machines. The image of the mechanical or energy slave was an old one, commonly used in the mid-nineteenth century for any labor-saving device that could harness new fuels to replace the flesh-and-blood worker. A logical outgrowth of the emerging tendency to view the human body as a human motor, this equivalency between conscripted labor and mechanical power is preserved in the word *robot*.[19] The term *energy slave* had been revived in the 1940s by figures such as the American architect and futurist R. Buckminster Fuller, with a new emphasis on the quantification of energy consumption. The metaphor came, of course, laced with vicious historical and racialized baggage that they largely ignored.[20] In Saha's hands, though, it took on a new—if understated—resonance.

In the pages of *Science and Culture* Saha did not discuss caste, perhaps a strategic silence.[21] Yet the manpower discussed in his 1944 article and the life of "indescribable wretchedness" that went along with it were inseparable from the brutal inequities of the caste system. Saha valorized work—"wealth can only be created by doing work"—and had earlier argued that one of the caste system's most pernicious consequences was its denigration of manual labor, thus distancing upper-caste students from the practical mechanics that had powered scientific innovation in the West.[22] Nonetheless, he passionately believed that the "future belongs to those who know how to use machines as slaves and not ask human and animal muscles to bear the strains which machine can bear."[23] Several scholars have argued that thermodynamics provided the ideology for the colonial-capitalist exploitation of raced laboring bodies, seeing in the concept of the energy slave a "techno-fundamentalist fantasy" that serves "modernity's privileged classes" by concealing exploitation and injustice within and across borders.[24] Just as historians have drawn provocative connections between steam power and the abolition of slavery in the United States, though, for Saha modern energy and mechanization were a force for emancipation.[25] In this optimistic subaltern perspective, it was not the soft-handed elites whom the machines would serve, but the vast body of the population who would finally escape exhausting toil. Saha was in some ways the early ancestor of twenty-first-century socialists who see revolutionary potential in the next generation of automation.[26] Such a view was still a fantasy, seeing in the new energy technologies a tabula rasa, the means to make a sharp break with past society. It willfully overlooked the continuation of caste and class privilege and subaltern suffering in new forms within the modern energy economy, a blind spot that would come back to haunt Saha.

In its full-throated endorsement of the machine age, this technophilic pragmatism was a ferocious attack on another leading vision for India's future: the back-to-the-village traditionalism of M. K. Gandhi. Fundamental to the Mahatma's critique of Western industrialism was the simple technology of the charkha (spinning wheel) as the vehicle of economic and spiritual salvation.[27] His followers heralded this as nothing less than an energy revolution. India's mass of underemployed laborers were remarkably effective "engines" fueled by food ultimately derived from the solar bounty of the tropics, the American pacificist Richard Gregg wrote, and the decentralized and frugal "food-man-charkha combination" would thus "save a vast existing waste of solar energy."[28] Saha regarded such views as at best foolishly nostalgic, at worst neocolonialist. In contrast to Japan, which "threw herself, with all her youthful energy into developing her natural resources in minerals and power" (which is discussed in chapter 3 of this volume), he scornfully wrote, "India, as well as China, burdened with her great past looks back with longing lingering eyes to the vanished village economy, to the cult of the spinning wheel and the bullock-cart."[29] In this way, Saha charged, Gandhians provided the British Raj with a "strange ally" in perpetuating the colony's scientific and industrial stagnation.[30] The past held little charm.

STATE POWER

To accelerate the essential energy revolution and thereby improve "the material condition of the average Indian," Saha called for central planning. Under British rule, electricity was almost entirely a provincial concern. He regarded this decentralization as an abdication of responsibility, not least because Britain itself, long something of an electrical laggard in Western terms, had at long last in 1926 begun to construct its own national grid.[31] While some Indian provinces had flourished (he singled out the progressive princely state of Mysore and the entrepreneurial Tata dynasty that supplied power to Bombay), many others merely "talked vociferously about industrialization" with little more in mind than "the manufacture of tooth-pastes and matchsticks."[32] Worse, as he had written earlier in the journal, private power companies in cities like Calcutta and Allahabad were "plain and simple *profiteers*." Civilized countries treated electricity as a public utility, based on the underlying idea that "electricity should be supplied at as cheap rate as possible to the public for all purposes, and as easy and uninterrupted supply of it should be secured just like the supply of water to a big city."[33] It was a powerful statement of the emergent notion that electricity was an essential public service and extraordinarily precocious, given that power still reached only a tiny minority of Indian households.

Saha's espousal of energy modernization and centralized state ownership was echoed by another sometime power planner and leading critic of the caste system, the great B. R. Ambedkar. The main drafter of India's constitution and the leading representative of the country's Untouchable community, Ambedkar had also chaired the colonial government's wartime committee on power. Despite Saha's and Ambedkar's pleas during the constitutional debates of the late 1940s, the principal responsibility for implementing power policy remained in the hands of subnational governments. The pair were prescient: this federal solution eventually led to a fateful politicization of electricity tariffs in the hands of subnational governments, draining Indian utilities of the resources for long-term electric expansion while incentivizing the pell-mell extraction of precious groundwater for agricultural irrigation in ways that now court ecological crisis.[34] Well into the twenty-first century, the virtual bankruptcy of subnational utilities meant that electrification for all would be sacrificed to the interests of the well-connected few.

Saha was more successful in advocating for his favored energy source. Today India is the world's second-largest consumer of coal behind China and thus often regarded as a "coal nation."[35] Under the Raj, though (and indeed up to the nationalization of coal in the 1970s), mining remained privately controlled, haphazard, and wildly inefficient. During World War II, shortages of coal and electricity became so severe in the eastern coal belt that the government was forced to relax its formal prohibition on employing women underground. In 1943, though, coal production actually fell, and railways and mills ground to a halt. It was a staggering indictment of an imperial regime whose own economic rise had self-evidently owed so much to cheap and abundant coal. Though this underproduction was in fact the result of misgovernance rather than geology, Saha concluded that India was "markedly deficient in coal" and that this "exhaustible national resource" must be therefore conserved for other essential purposes, such as chemistry.

Instead, Saha called for the construction of large multipurpose dams, which could provide not just hydroelectricity but also flood control, irrigation, navigation, malaria control, and soil conservation. This was a long-standing concern: he had carried out successful fundraising for flood relief efforts since 1923 and became so interested in "River Physics" that he constructed a model riverbed in the college corridor next to his office.[36] Only months earlier, in July 1943, the Damodar, India's "river of sorrow," had again burst its banks, cutting Calcutta off from the eastern battlefields while worsening the terrible man-made famine that would eventually leave an estimated three million Bengalis dead. It was a turning point in the delegitimization of the colonial regime, confirming to many Indian nationalists the importance of state action in the name of poverty alleviation.[37]

Saha was serving on the colonial committee established to enquire into the flood and used the opportunity to advocate for river valley development as the silver bullet. His piece cited two apparently very different models: the Tennessee Valley Authority in the United States, a New Deal megaproject he had long admired, and the great dam on the Dnieper in the Soviet Union, the product of Lenin's famous definition of communism as "soviet power plus electrification" (though ironically built with American equipment and expertise). Presaging India's complex positioning in the coming Cold War,

Saha himself was a Soviet sympathizer, while conceding that a "form of controlled capitalism, in which the profit motive shall be subordinated to the ideal of social welfare and service, will best suit Indian conditions and culture."[38] In any case, as Carl Schmitt had noted in 1923, ideological enemies converged almost entirely on the energy question: "The big industrialist has no other ideal than that of Lenin—an 'electrified earth.' They disagree essentially only on the correct method of electrification."[39] Cold War lending would bear out this convergence, with India successfully securing funds and other support from both sides for coal plants and big dams alike.[40]

Saha shared the ecological optimism of the two great powers. Modern scientific discoveries had "so far increased man's power over Nature" that the world could look forward to a future of "plenty for all," if only rulers could see beyond their lust for exploitation and see like "the scientific man." His brainchild, the vast multipurpose river valley development project of the Damodar Valley Corporation (DVC), would become the first of postcolonial India's grand energy projects. Though designed to be run by central technocrats (Saha's scientifically minded men), in practice it quickly ran into political problems: the humanitarian question of the thousands displaced by its reservoirs, endless squabbles between state governments for primacy, and populist calls for giving away its irrigation waters for free.[41] At the same time, Saha failed to recognize that big dams destroy whole ecosystems and release vast quantities of methane, even while the more-than-human forces of erosion and siltation gradually undermine their energetic potential. Many other mega-dams nonetheless followed the DVC. Decades later, the backlash against such projects would eventually play a crucial role in galvanizing India's noisy and multifaceted environmental movement.

If the ecological repercussions of rising energy consumption remained largely invisible to Saha, as for many of his contemporaries, his accounting nonetheless lives on within the environmental movement in a very different fashion. Saha's energy-index used the low electricity consumption *per head* of India's growing population as an indicator of backwardness and, by extension, evidence of colonial misrule. More recently, Indians have repurposed similar per capita figures as a new metric of global climate justice. Echoing an older Malthusian obsession, India's critics in the rich world point to a favorite figure to argue that the country's energy hunger is a planetary threat: the overall national emissions produced by its huge population, especially as economic growth picked up speed through the 1980s. Indians have rejected such aggregates as an "example of environmental colonialism" and instead placed the obligation for climate action squarely on the shoulders of wealthy countries.[42] Pairing a similar per capita metric with a powerful spatial metaphor, in 1991 the country's most celebrated environmentalists argued for conceptualizing global warming in terms of "sink space": "the oceanic and tropospheric sinks" were "a common heritage of humankind" and thus a global commons that should be "shared equally on a per capita basis."[43] Any attempt to force India to curtail its energy growth was, in the words of the chief economic advisor to the government of India, "carbon imperialism."[44] Saha used his per capita calculations to urge the state to take a leading role in accelerating the belated energy revolution in the subcontinent. In the age of climate change, India's low per capita energy consumption has become a way for its leaders to argue for the continued right to develop, even as the atmospheric global "carbon space" (the metaphor that has won out) for energy-hungry industrialization shrinks to dangerous levels. For Saha and for contemporary environmentalists in their very different ways, the lessons (and mistakes) of Western development could be learned from without compromising India's own projects of emancipation.

Despite his intellectual brilliance and institution-building efforts, Saha found himself increasingly on the outside of a regime still dominated by the wealthy upper castes. After a tour of European and American laboratories in 1936–37, he initiated the country's earliest serious research into nuclear physics, and in 1944–45 led another delegation to examine Western developments in the geosciences. Yet after independence in 1947 he was cut off from the atomic energy establishment in favor of his old upper-caste rivals, just as the new country's loss of oilfields in Burma and Pakistan intensified the urgency of finding a new answer to India's energy needs. As the postcolonial state became ever more enamored of expensive and secretive nuclear projects using imported components and expertise, Saha instead urged his compatriots to fund basic science and domestic equipment so that the nation's technological development might become self-sufficient. He was defeated, many commentators writing his misgivings off as the complaints of a bitter also-ran or the inevitable product of his impoverished background.[45] Saha had been too optimistic that the new energy technologies would break the grip of the old elites, too optimistic that science could stand above the messy realities of politics.

Increasingly disillusioned, Saha became an independent member of parliament and an incisive critic on a wide range of technical and social issues. In February 1956 he died suddenly on his way to the offices of the Planning Commission, all too aware of the widening gulf between the country's trajectory and his vision of social justice powered by modern science and newly abundant energy. His last-ever article, published in *Nature* three months after his death, was a careful calculation of the country's fuel usage a decade into its great postcolonial experiment: oil and hydroelectricity, 5.6 percent; coal, 28.0 percent; cowdung and firewood, an overwhelming 66.4 percent.[46] The modern energy revolution remained only a thin topsoil on an economic bedrock of traditional fuels and the muscle power of its poorest members. For most Indians the age of electricity had yet to dawn, even as Saha's beloved project of energy modernization began to choke the country's rivers and skies.

MEGHNAD SAHA
"India's Need for Power Development" (1944)*

The modern age is fundamentally different from medieval times, and there are many ways of describing the difference. We sometimes say that we are in the age of steam or electricity. Another writer describes the modern age as

---

*Saha, "India's Need for Power Development," *Science and Culture*, 10, no. 6 (1944): unpaginated (selected extracts). Based on a lecture delivered by Prof. M. N. Saha before the Calcutta Rotary Club on July 10, 1944. Permission for reprinting granted by the Indian Science News Association.

"neotechnic" (production in factories by industrialists with the aid of large machineries) as distinguished from "paleotechnic" (production in cottages by artisans working with small tools).* The third definition, though accurate, is pedantic. But probably nothing brings out the characteristic difference so vividly as the *energy-index*—by which term we mean "the output of energy or work per capita of the population per year from all sources." Probably it will surprise some of our readers to learn that the energy-index† in medieval times was not more than 80, while in modern times it is nearly 2,000 in a moderately advanced country and in the neighbourhood of 3,000 in some of the go-ahead countries like U.S.A. or Canada.

What do these figures indicate? It shows that an average citizen of an

---

*[*Ed.*: This slightly misleading summary refers to the work of Scottish social evolutionist and urban planner Patrick Geddes (1854–1932), who drew a distinction between the earlier "paleotechnic" period of coal-powered industrialization, which he associated with waste and pollution, and the finer "neotechnic" period of electricity, which promised to restore the environment and rural community; see especially Patrick Geddes, *Cities in Evolution: An Introduction To The Town Planning Movement and the Study of Civics* (London: Williams & Norgate, 1915). Geddes had carried out exhaustive studies of eighteen Indian cities in the 1910s and so was well known in the subcontinent. He was also a major influence on the American philosopher of technology Lewis Mumford (1895–1990), who picked up the paleotechnic-neotechnic frame in his 1934 classic *Technics and Civilization* to condemn "carboniferous capitalism." Geddes continued to influence Indian scholarship into the 1940s, most notably the evolutionary approach and ecological regionalism of sociologist Radhakamal Mukerjee. Lewis Mumford, *Technics and Civilization* (New York: Harcourt, Brace, 1934).]

†Note for the non-technical readers: The "unit of energy" used here is the "kilowatt-hour," i.e., the work done by a machine having a power of one kilowatt for one hour. This is equal to one and one-third horse-power hour. The man-power hour is just one tenth of one horse-power hour, i.e., an average adult working for 8 hours produces only 6/10th of a unit (kilowatt-hour). The price of a kwh, charged by an electrical company for industrial use is about an anna, so that a man working for 8 hours does work, whose normal value is 2⅓ pice [*Ed.*: 1 rupee = 16 annas = 64 pice]. But generally man's work, even if it be entirely manual, is intelligent, hence the value is higher. For very unintelligent or mechanical work, like spinning or digging or drawing water, the rating will be lower, say at twice the minimum.

"India's Need for Power Development" (1944)

advanced country, like Sweden, is today nearly 25 times richer than he was in medieval times, for wealth is directly proportional to the output of work. True, the total national wealth in most countries is not equitably distributed, but that is a different topic. If the country as a whole gets richer everybody shares in the benefit, though there is no denying the fact that an equitable distribution is far more beneficial and desirable. But the question at any rate is of secondary importance, for unless there be production there would be nothing to distribute. Countries, like Iran or China, which have till recently clung to medievalism, can never attain in spite of the traditions of a great past, or prevalence of saints or wise men, the prosperity, or material happiness of modern Sweden.

How has this profound revolution been brought about?

In medieval times, work was done almost entirely by manual labour aided by animal power (cattle and horse) and to a small extent by powers of nature (wind power or water power as used in water wheels first invented in Iran). In modern times work is done by power derived from coal (steam engines) and oil (oil engines) and by electricity generated from coal (thermal stations), oil (Diesel engine) or water power (hydro-electricity). For a country it is now possible to calculate, almost with mathematical precision, the whole amount of energy output from coal or oil consumed for this purpose, or from the records of electrical power companies. The total energy output divided by the population gives us the energy-index.

Countries, like Sweden and Switzerland, which have very little coal and oil have to depend on electrical power generated almost entirely from water. In 1942, Sweden's output of work per capita per year was nearly 1,700 units. In a country where steam power is very largely used, such as England, calculations are more difficult to carry out. The per capita energy derived from steam and other sources has been calculated to be about 1,400 units (this figure is given with a certain amount of reserve) so that the energy index for the U.K. amounted to about 2,100 units. The figures for Canada and U.S.A. are higher, probably in the neighbourhood of 3,000 units. The U.S.A. and Canada have enormous resources of water power and have developed a very large percentage of that, with the result that they are now ahead of the whole world in the production of work and in material prosperity.

Where do man power and animal power come in this picture? A man's output of work in the whole working year of 300 days is only .6 × 300 = 180 units, and since we have to leave out old men, children, most of women and the idle rich for purposes of productive work, we can take the active worker as one in three. The average annual output by man power is therefore only 60. The output from animal power could not have been more than 20, so that in medieval times, the energy-index could not have been larger than 80. The output of work by man's physical exertions can now be entirely neglected in comparison to the energy output from other sources; his work is now mainly directive.

If we arrange the countries of the World according to the energy-index, at the bottom amongst the countries which claim to be civilized would be found India and China, and pre-Bolshevik Russia whose position in 1918 was no better than that of India in 1944. India's energy-index in 1944 cannot be much larger than 90 units even on the most liberal estimate; this is made up of 60 units derived from manual labour, 15 from animals, 9 from electricity in 1942 (the total production was 3,500 million units) and about 10 units from steam and oil—the exact figures being not available. So the energy-index cannot be larger than 90 units, as against the Euro-American figure of 2,000 or more.

The average annual income of an Indian was calculated by the National Planning Committee to be Rs. 65. The average of a Swedish was about 1,800 Kroners in 1938, or about Rs. 1,300. He earns 20 times more than an Indian. The reason is quite clear; he uses about 20 times more energy. In fact we can assume that the average income of the citizen of any country from all sources would be directly proportional to the energy-index.

We can put the facts in more figurative language. In Europe and America forces of nature have been harnessed so effectively that a European or an American has literally 10–15 slaves working constantly for him. In India, the number is 1½, 1 being himself and ½ derived from the harnessing of forces of nature and from animal power. We may say that in India the energy output is equivalent to 1½ slaves, of which 1 is the man himself and ½ derived from other sources. This is probably just equal to half a donkey-power.

"India's Need for Power Development" (1944)

UTILIZATION OF ENERGY

[. . .] It is this large production with the help of natural power which has enabled men in advanced countries to attain a standard of living far beyond the dreams of medieval philosophers, a fact which is reflected by the rise in average longevity from 25 in about 1870 to over 55 in 1938 in Europe and America.* [. . .] It is known to everybody that 90 per cent of the people of India still lives in the sixteenth century—a life of chronic malnutrition amounting in times to famine, disease and indescribable wretchedness. The amenities of modern life are available only to a small fraction of people living in the cities. Yet India has all the resources in power and materials which, if they were properly exploited, could raise her to the standard of the average European countries. The paradox is, as Dr Vera Anstey succinctly puts it, that the soil of India is extremely rich, but her people is extremely poor.†

In the past, there has been much loose thinking regarding the methods by which the material condition of the average Indian can be improved, but it is only recently that there has been some objective thinking. This first step towards this was taken by the National Planning Committee of the Indian national Congress which sat during the years 1939 to 1941 at Bombay under the Chairmanship of Pt. Jawaharlal Nehru. It is pleasing to note that the Committee in its work secured ready and helpful cooperation from the Provincial Governments including those, like Bengal, which were not run by Congress Ministry. This shows the popularity of the idea of national planning. The Committee found that if the material condition of the average Indian were to be substantially raised within the next ten years, her production of energy should be raised by 40,000 million units per year (above the present figure). This may sound to be a very large figure, but a little reflection will show that the estimate is extremely modest. Even Mexico which is considered to be a rather backward country produces 180 units of electrical

---

*According to a report in *Science*, May 19, 1944, the longevity in the U.S.A. reached a peak of 63.82 years in 1942.

†[*Ed.:* Vera Anstey (1889–1976) taught economics to a generation of South Asian students at the London School of Economics.]

energy per capita per year contrasted to India's nine in spite of the chronic revolutions for which she had acquired a notoriety. The National Planning Committee's figure gives only 100 units per head, which is 1/20th of the per capita production in the U.S.A.

This energy can be generated by burning 20 million tons of coal, but India is markedly deficient in coal, and as an exhaustible national resource on which many essential needs of the nation depend it should be conserved for other purposes as far as practicable. It is desirable that most of this energy should come from electricity, derived from hydro-electric sources* which exist in plenty. Further only a small part of India has coal; other parts, such as the Punjab, Bombay and the whole of South India, have to depend on longhaul coal. It is not only desirable, but almost imperative that the water power resources of these parts be properly developed. The hydro-electric survey commission should now be revived, if we really wish to pass out of the present axis combination of poverty, disease, and malnutrition.

In the reforms of 1933, development of electrical power was made a "Provincial Obligation." The Government of India thus divested itself by one stroke of pen of one of the most important obligations of Central Governments of all countries, and the befuddled Provincial Governments mostly talked vociferously about industrialization which sometimes in their minds was identical with the manufacture of tooth-pastes and matchsticks without achieving anything, and with one or two exceptions did practically nothing useful. [...]

In India, we have many populous river valleys which are subject to vio-

---

*India's hydel [a portmanteau of "hydro" and "electric"] power resources were estimated or rather guessed by [J. W.] Mear[e]s in 1922 to be 20 million kilowatts, so that even on this estimate up to this time, barely 1 per cent of the resources has been harnessed. But Mear[e]s' figure is most probably glaring underestimation, for he was asked to do things hurriedly and he had neither the time, nor the proper resource to make an accurate survey. A parallel is afforded by Soviet Russia which before 1918 was reported to have hydel power resources amounting to 20 million kilowatts; but accurate survey carried out by the Soviets showed that the figure was 280 million kilowatts or 14 times higher. After the last War the Government of India had a plan for larges-cale development of hydel power in India, but this was dropped in 1923.

"India's Need for Power Development" (1944)

lent floods and which, on account of thoughtless handling in the past, have gone down in prosperity and public health. The most glaring examples are the Damodar Valley in Bengal and the Mahanadi Valley which is practically the whole of Orissa. How these rivers can be turned into beneficial agencies is afforded by the example of the Tennessee Valley in the U.S.A. This area, comprising about 40,000 square miles was going down continuously, and in 1931, it was estimated that for a large part the income had dwindled to 100 dollars per capita in the year. In 1933, due to the personal initiative of President Roosevelt, the Tennessee Valley Authority (shortly called TVA) was created and within ten years the whole face of the country has been changed. The core of the whole work has been the construction of about 20 multipurpose dams built for flood prevention, irrigation and soil conservation, navigation, and power generation. In 1943 according to the annual report of the Authority 9,000 million units of electricity have been generated, (a summary of the report appears elsewhere in this issue), and about a stretch of 640 miles of the river had been rendered navigable. Agriculture and industry have been in a flourishing state. Jacks and Whyte put it:

"Nature and vested interests turned a virgin country into a wilderness; but nature and man have again combined to make it a smiling garden."

It has been shown that the Damodar River Valley which is now notorious for its floods and malaria can be treated in the same way as the Tennessee and be turned into a beneficial agency yielding nearly 1,200 million units of hydel energy. [. . .]*

We have tried to impress upon the reader the basic fact that the root cause of poverty of India is the hopelessly inadequate use of natural power. This is self-evident. Because, wealth can only be created by doing work. And, the work necessary for creating wealth of [an] amount approaching that of the western countries cannot be done by man and animal power alone as is often held in this country. India must, therefore, develop and utilize to a vastly greater extent her natural power resources than she has been doing till now. Lenin, the Father of new Russia, realized the supreme necessity of develop-

---

*"Hydel" is a portmanteau meaning hydroelectric.

ing power. One of his first acts on assumption of power was the appointment of a commission under Prof. Krzhizhanovsky [Saha refers to Krzhizhanovskii; see Russ, ch. 6] to enquire into the power resources of the country. In fact, Russia began the development of power long before she launched her Five Year Plans. The Great Dnieper Dam was nearing completion when she started her First Five Year Plan in 1928. It is, therefore, imperative that the Central Government should at once set up a committee to make a thorough survey of the power resources of the country as a whole and establish suitable educational institutions to train a proper personnel. The survey is of basic importance and should be started immediately without waiting for the full formulation of the reconstruction plans.

[. . .] Would India enter on a period of planned development and prosperity, or would all the mistakes which were committed after the first World War be repeated?

6 The Red Thread to Socialism

Gleb M. Krzhizhanovskii's "Energetics and Socialist Reconstruction"

DANIELA RUSS

IN RUSSIAN, LIKE MANY European languages, the metaphor of a red thread symbolizes both orientation, as in Ariadne's mythic thread that led Theseus through the labyrinth, and identification, as in the red thread woven into the British Royal Navy's ropes to impede theft. The expression stems from Johann Wolfgang von Goethe's *Elective Affinities* (1809), where he likened the relation between a text and its leitmotif to the Royal Navy's ropes that "are so twisted that a red thread runs through them from end to end, which cannot be extracted without undoing the whole; and by which the smallest pieces may be recognized as belonging to the crown."[1] A red thread ties a complex unit together, gives it structure, and makes each part recognizable as parts of the whole. This unit can be a story—or an entire economy, as Gleb Krzhizhanovskii, the author of the text presented here for the first time in translation, suggests. As head of the Soviet Union's famous electrification commission (GOELRO) and planning committee (Gosplan), Krzhizhanovskii thought of energetics as a 'red thread' that would lead the Russian economy into a socialist future.[2] Not only would energy politics be able to guide socialist transformation and hold the economic whole together, but socialist states would, he argued, become distinguished by the organization of their energy economies. When Krzhizhanovskii left the Soviet

Union's State Planning Commission (Gosplan) in 1930, these hopes were frustrated: Stalin's program of accelerated industrialization had just begun, and the targets of the first Five-Year Plan (*piatiletka*) had been repeatedly increased until they ceased to make sense. Not only were the intended output levels for secondary resources too high, so too were the energy targets, which Krzhizhanovskii saw as the very "backbone" of the Socialist economy.[3]

In the early Soviet Union, energy and economic planning had been tightly linked. Vladimir Lenin, chairman of the Council of People's Commissars of the Russian Soviet Republic, affirmed his commitment to electrification in the now famous dictum that "communism is Soviet power plus electrification of the entire country."[4] Lenin also made sure that electric engineers who had worked in the electrification commission, such as his old friend Krzhizhanovskii, were promoted to influential positions within the state's economic planning organization. There, Krzhizhanovskii had forcefully promoted his second "general plan" for years, as a follow-up to GOELRO. In "Energetics and Socialist Reconstruction," he put forth the argument that Socialist transformation required an energy analysis of economic processes and that planning should be guided by its outcomes. However, this visionary project proved of no avail. When Lenin died in 1924 and a power struggle within the party began, the electric engineers' situation became more vulnerable. That the main agency for electrification (Glavelektro) was headed by Trotsky, Stalin's fiercest opponent, also proved consequential.[5] Opposition to scientific and energetic planning was growing among the highest party officials. At the time this text was written in 1929, the struggle for Krzhizhanovskii's plan had already been lost. "We were born too early," he later commented to a fellow energy engineer.[6]

It is ironic that today's Russia is notable for its energy economy, but not in the way Krzhizhanovskii had hoped. Like many Soviet engineers in the 1920s and 1930s, he saw petroleum as a transitory fuel and instead supported a broad approach to electrification based on locally available sources such as peat, coal, and flowing water and then, eventually, solar and nuclear power. While the Bolsheviks exported oil for hard currency, its imperial legacy, use in individual transport, and value as industrial feedstock made it a less attractive basis for the domestic Soviet energy economy, which mainly ran on coal. This view began to change slowly when the Soviet Union was forced into a war against Germany, a conflict it was unable to fight without kerosene-fueled airplanes and diesel-fueled tanks.⁷ By the late 1950s, the Soviet Union had embarked upon the well-trodden path towards an oil-based economy. After the pipeline boom of the 1960s and 1970s, oil and gas flowed into the West in large volumes and hydrocarbon revenues provided a vital injection into a floundering Soviet economy.⁸ The runaway costs of oil and gas extraction, which required drilling deep into Siberian permafrost, can partly be blamed for the late Soviet Union's economic malaise.⁹ Today, Russia is a fossil-fuel rent-based economy; for the last decade oil and gas tax revenues made up around 50 percent of its annual budget, a situation that makes decarbonization extremely unlikely.¹⁰

Amid a flood of scholarship on the Anthropocene, most of which has focused on capitalist and imperialist expansion, histories of socialist and post-colonial energy use have shown that for some countries the building and expansion of fossil economies was once a state-led project of emancipation.¹¹ While these projects read differently from our warming present, they remind us of the deeper meaning once attached to the collective organization of energy. Recently, visions of an electrified modernity have returned in the wake of the politics of decarbonization.¹² The project to "electrify everything" from heating to transport does not only revive the idea of electricity as a rationalizing force. As more and more sectors become coupled together, it also poses the question of the social coordination of energy use and the need for fine-grained energy accounting. As such, the current historical juncture may be a good time to remember Krzhizhanovskii's somewhat forgotten energy-economic thinking.

A convinced Bolshevik and broadly trained engineer, Krzhizhanovskii blended the international electrotechnical discourse on conservation and efficiency with Lenin's theory of imperialism and Friedrich Engels's nature-dialectics. The result was an engineer's diagnosis of how the capitalist economy had reached a point of energetic impasse and a plan for how the Soviet Union could learn from this and liberate itself on the basis of the most progressive force of production—electricity—and the social relations it forged. According to Krzhizhanovskii, this would require energetic planning at the level of the state. In the Soviet economy "energy," understood as the relations between different forms of power, such as oil, coal, heat, electricity, etc., became a political problem long before the crises of the 1970s.¹³

As a student at the prestigious Technological Institute in St. Petersburg in the 1890s, Krzhizhanovskii received wide-ranging training in mathematics, engineering, physics, and chemistry, graduating as "engineer-technolog" with a major in chemistry. At the time, the Institute's student body was infamous for its rebellious energy and well-known for its social-democratic rather than "Narodnik" views.[14] Krzhizhanovskii was part of a leftist reading group, which Lenin soon joined and which became one seed of the Russian Social Democratic Party.[15] A schism in Russian Marxism had occurred in the late nineteenth century, when Georgi Plekhanov and Lenin began to distinguish their focus on large-scale industrialization along a Western ("social-democratic") path from other Russian Marxists (dubbed "Narodniki") who were searching for a distinctly Russian path to socialism based on the peasantry and village communes.[16] Krzhizhanovskii's worldview developed amid these debates. While there are some early indications that he had been interested in the Narodnikis' plans to improve small village economies in the 1890s, by the 1920s, Krzhizhanovskii was convinced the only way for the Soviet Union to survive was large-scale and high-tech industrialization.[17]

World War I proved crucial to this change of perspective. In the view of many Bolsheviks, the war had smashed any remaining hopes to develop a viable Socialist state based on rural craftsmanship of the kind that had developed in Imperial Russia. "The war taught us much," noted Lenin in 1918, ". . . but especially the fact that those who have the best technology, organization, discipline and the best machines emerge on top; it is this the war has taught us, and it is a good thing it has taught us. It is essential to learn that without machines, without discipline, it is impossible to live in modern society. It is necessary to master the highest technology or be crushed."[18] Lenin's attempt to win over the technical intelligentsia to the Bolshevik cause, his embrace of scientific management, or Taylorism, and scientific planning has to be seen in this context.[19] Working alongside non-Bolshevik technical experts in GOELRO and Gosplan, Krzhizhanovskii strongly supported Lenin's policy to involve the "spetsy" (specialists) into the state apparatus, which Stalin was deeply skeptical of. Because of this disagreement, in 1930, Krzhizhanovskii found himself sidelined into the Russian Academy of Science, where he became director of the Institute of Energetics.

Krzhizhanovskii was one of the last first-generation Bolsheviks ousted from power in the late 1920s, as Stalin tightened his rule over the entire state apparatus and curbed the influence of technical experts.[20] At a time when Gosplan's authority was already waning, the essay "Energetics and Socialist

Reconstruction" was Krzhizhanovskii's response to Stalin and his confidant and second in command Viacheslav Molotov's industrialization policy, which was based on the development of heavy industry (coal, steel, and fabricating productive machinery). Published in Gosplan's monthly journal *Planning Economy* in 1929, the text was carefully crafted to avoid direct confrontation with the regime, which could prove deadly, while also summoning the legacies of Lenin and Engels to aid his cause. Below many qualifications and nuances lay the clear argument that mere industrialization—that is, a focus on heavy industry—would not lead to a truly Socialist economy. To achieve this, a full transformation of the "technical-economic basis" was required, a remaking of the structures along which energy and matter flowed.

It has rarely been noted that this was more than a power struggle. Few historians saw much of a difference between the two proposals in terms of concrete policies; after all, Stalin and Molotov never planned to give up on electrification.[21] Electrification, in their view, would follow the needs of older industries—metallurgy, railroads—rather than leading the way towards new ones.[22] For Krzhizhanovskii, however, there was more at stake: the point he tried to repeatedly make was that creating an energetically optimized economy was not a question of introducing any single measure or technology.[23] It was not as simple as encouraging full electrification; a more holistic energy analysis of economic processes was needed: Krzhizhanovskii wanted a system that accounted for the material-energetic efficiency of the *entire economic organism* and did so from an institutionalized position—a "commanding height" from which the economy could be constantly evaluated, organized, and developed along energetically optimal principles.[24] It was this idea of central energetic planning, which would have secured great influence for energy engineers, that was lost in Krzhizhanovskii's defeat.

In the article, Krzhizhanovskii framed the question how the Soviet Union should industrialize in the schematic "Hegelian" form typical of dialectical materialists: How could *quantitative* economic processes (higher productivity, more powerful machinery, faster transport, etc.) be transformed into *qualitative* changes—that is, a change in the structures of economic reproduction and in the social relations of production—so that a truly different, Socialist economy can emerge from it? He conceded at the outset that his proposed "new technological basis" and Molotov's development of heavy industry were somewhat aligned but emphasized that industrial progress alone was an "insufficient condition for translating the quantitative economic changes into a socialist quality of the economy." At the critical juncture between economic restoration and socialist reconstruction, a more precise guide was needed. In Krzhizhanovskii's view, this required historical consciousness, a scientific analysis of past developments, and a sense for the current moment. In such analysis, Krzhizhanovskii explained, it would become clear that a focus on heavy industry was something characteristic of the industrialization of the past while electricity would instead lead into the future.

Like other Soviet engineers of the time, Krzhizhanovskii interpreted the contradiction between *forces* and *relations* of production in a distinctly energetic way. In the orthodox Marxist reading, *productive forces*—the resources, knowledge, and technologies applied to produce goods—would come into conflict with the *relations of production*—the social relations around production (i.e., wage labor, a certain division of labor, or property regimes). In the energetic reading of this contradiction put forth by Krzhizhanovskii and others, it was not any part of the productive forces but the prime mover of machinery from water to steam and electricity that had and would revolutionize production.[25] Citing Friedrich Engels,[26] Krzhizhanovskii saw electricity as a "rationalizing element" within a wider range of productive forces: Electricity functioned as a prime mover, transmission device, and instrument of labor all at once. It affected all transformations that took place in production, as well as the vital metabolism between nature and society.[27] Electricity's capacity to convert forms of energy into one another promised to eliminate the metabolic rift between the city and countryside, as industry (and human settlements) could be distributed more equally over space. Electricity could be used to transform matter, form chemical elements, and even produce minerals that had once been taken from the soil. This sparked hopes of a more harmonious relation between nature and society. With Socialist technology and particularly socialist forms of electricity use, as Krzhizhanovskii had contemplated in another text, "man finally senses [*nashchupyvat*[28]] the ways in which the powerful creation of his hands can be included in nature merely as an element that ennobles it."[29]

As a productive force, electricity could transform not only the composition of the Earth but also the fate of its workers: it could remake the relations of production. A popular slogan among Soviet engineers was that "capitalism is the age of the steam engine" and "socialism the age of electricity," an extension of Marx's "the hand-mill gives you society with the feudal lord;

the steam-mill society with the industrial capitalist."[30] Hidden in this was an argument about how a new society, new *relations of production*, would emerge from a particular technology. Within the factory, where an "autocracy of engineers" had formerly directed an "aristocracy of machinery" to subordinate living labor (the worker), whereas a socialized energetics based on electricity would instead allow "living labor to stand above a machine not in the form of an individual creator of this machine, an engineer, but in the form of a conscious human collective armed with the creative thought of centuries."[31]

Influenced by an electrotechnical paradigm that focused on interconnection and economies of scale, the GOELRO engineers had planned to create a state network of large regional stations ranging from 40 to 60 megawatts.[32] While this was a normal size for other countries, it amounted to a significant integration and concentration of power production in Russia, which was more used to small urban and industrial stations.[33] Unlike with petroleum lamps and automobiles, the generation of electricity for individual use was extremely expensive and inefficient, whereas if collectivized, it could become more economical than other forms of power. In this materialist and technologically determinist thinking, energetic optimization, via the coordination of sectors and industries, would create a new subject, conscious of her powerful, collective control over the forces of nature. In practice, however, large-scale electrification did not mold the Soviet population into a single human collective; cities and villages fought for more small-scale electrification and greater control over it.[34]

The hypothesis that there was a contradiction between energetic forces and social relations of production rested on a third element: the increasing concentration of productive power that took place under capitalism, which had been documented by Lenin and others. According to Krzhizhanovskii, this energetic contradiction consisted of the fact that the most progressive forces of production (electrification) demanded a collective form of organization that capitalist relations of production (relations of capital, property, and ownership) could realize only by undermining themselves. This could be seen from the anti-competitive forms, such as the trusts and cartels, that had shot up amid "free" competition. In *Imperialism: The Highest Stage of Capitalism* (1916), Lenin had measured the concentration of productive power not only in the number of an economy's enterprises and employees but also in the share of the total steam and electric power they employed.

He found that in Germany, "less than one hundredth of the total enterprises utilize *more than three-fourths* of the steam and electric power!"[35] The control over working nature was even more concentrated than that over laboring bodies, and to achieve it, companies resorted to material planning. Large electric or oil companies such as Standard Oil, General Electric, or the German Allgemeine Elektrizitätsgesellschaft (AEG) epitomized this consolidation of productive power, as their operations spanned continents and spread across various industries; a list of the AEG's worldwide subcompanies, from banking to batteries, already filled four full pages in the German social-democratic newspaper *Die Neue Zeit* in 1913.[36]

In the understanding of some Marxists, the size of trusts, their ability to restrict competition and to rationalize production within their operational jurisdictions, signaled the coming of more socialist forms of production. Rather than waiting for these economically rational organizations to form out of the chaos of competition, Krzhizhanovskii believed the Soviet Union could create them, as it had overcome the conflict among capitalists and between capitalists and workers. An enthusiast of descriptive geometry since his university days, he expressed the elimination of internal contradictions in a geometrical metaphor: if social forces were imagined as lines of force, socialism would mean to align the wills of the working people until they formed a "parallelism," one massive social force pulling in the same direction.[37] Among the "great works" such a force could realize were "energy-industrial complexes," large productive units in which energy and material flows would be planned in a manner that integrated industries, households, and transport. As an example, Krzhizhanovskii mentioned DneproGES in Zaporishia, today's Ukraine, which was one of the largest hydroelectric dams in the world when it was finished in 1932. However, for those who had built the facility in harsh conditions, particularly after Stalin's accelerated industrialization and agricultural collectivization had increased general hunger and hardship, Krzhizhanonvskii's vision of Soviet people forming a single world-making force must have rung hollow.[38]

What makes Krzhizhanovskii's thought surprisingly topical today is his belief that capitalism had run against an energetic impasse that it itself had brought about. In his understanding of the historical development of productive forces, the evolution of steam-powered industrial capitalism and the metabolism it had depended upon had come "to an end at this energetic stage of the productive forces." His intervention advocated for planning based on overarching energetic principles at the level of the state. It was at Gosplan, the "commanding height" of the economy, where flows of matter and energy should be tracked and optimized, creating the material precondition for true socialism. While Krzhizhanovskii can be read as putting forth a linear theory of history which culminated in an electrified economy, he added the remarkable possibility that such an economy could have a different social form. The "strictly centralized economic system" could eventually give way to "the flowering of free economic communes, which, perhaps, will rely on entirely new material bases," such as nuclear power. Whether, in alluding to a kind of atomic communalism, he was thinking of his earlier work on the village commune and he truly hoped to overcome central planning or whether this was meant to win over skeptics by depicting centralization as temporary remains uncertain. Within the metaphor chosen by him, however, this openness of the future makes sense. A red thread, after all, is not a map but a means of navigation where the digressions and meandering of the route become visible only upon travelling along it.

GLEB M. KRZHIZHANOVSKII
"Energetics and Socialist Reconstruction" (1929)*

INDUSTRIALIZATION AND ENERGETICS

The period of socialist transformation, also known as the period of socialist reconstruction, pushes questions concerning the qualitative nature of our economy [*khoziaistvovaniia*, doing-economy] to the forefront, for the simple reason that any effort towards a socialist transformation presupposes the

---

*Gleb M. Krzhizhanovskii, "Energetics and Socialist Reconstruction," *Planovoe Khoziaistvo*, no. 1 (1929): 7–53 (selected extracts). Translated by Daniela Russ. Italics in the original.

Gleb M. Krzhizhanovskii

transformation of a range of *quantitative* processes into those of a *qualitative* character.*

This work has an ideological aspect: our basic targets, which constitute something like the pillars of our economic construction, must also be newly defined. So far, a broad range of terms have been used to refer to the transformation of these basic structures. We often hear, for instance, that the fundamental task is to provide our economy with a *new technological basis*, which includes electrification as well as other characteristic achievements of postwar technology. In addition, we are also considering rebuilding our economy on the basis of *heavy machine industry* [*krupnoi mashinoi industrii*] and the *industrialization* of the country. Both of these objectives ultimately aim to provide the economy with a new technological basis. However, concerning the guiding structures of our economic construction, the period of socialist reconstruction will require a larger degree of precision and clarity than was permissible in the period of basic restoration.

The term "industrialization" is the first, absolutely correct, and broadest generalization of the results we arrive at after studying the economic development of countries that are ahead of us in technical and economic terms. [...]

However, industrial progress is a necessary but insufficient condition for translating quantitative economic changes into a socialist quality of the economy. The history of industrialization presents a series of such stages.

---

*In the earlier version, Krzhizhanovskii begins by saying that the period of recovery was marked by an "economic drift" or "digression" (*samotek*). The word *drift*—an uncontrolled movement—probably refers to the period of the New Economic Policy (1922–1928), in which a free market was temporarily allowed to form in certain parts of the Soviet economy to overcome the economic malaise after the First World War and ensuing civil war. In Krzhizhanovskii's understanding, this period of "basic restoration" served to revive pre-war industrial capacity (in quantitative terms) but did not change the Soviet economy into a qualitatively different, socialist economy. The subject of his intervention is the following period of "socialist reconstruction" or "transformation," which would be marked by a greater significance of planning.

"Energetics and Socialist Reconstruction" (1929)

The steam engine—whose birthplace was the kingdom of coal and metal—ushered in the true industrial era and laid the foundation for heavy industry. Yet the interconnectedness of technology and its continuous progress, influenced by an array of economic and historical occurrences, leads to a variegated composition of the complex whole that we call industry. At the beginning of the nineties, [Friedrich] Engels wrote to Nikolaison:

> Industrial production at the present time certainly means large-scale industry—the use of steam, electricity, self-acting spinning and weaving machines, and, finally, the machine production of machines themselves. From the moment Russia introduced railways, the introduction of all these new means of production had been decided. You *must* be able to *build* and fix your own locomotives, wagons, railways, etc., and to do so cheaply, you must be able to build all the items you need for producing and fixing them at home. From the moment military affairs became one of the branches of large-scale industry[, . . .] large-scale industry, without which none of these things can be manufactured, has become a political necessity for you as well. All these things cannot be manufactured without highly developed metallurgical production, and metallurgical production cannot develop without the corresponding development of all other branches of manufacture and, in particular, textiles.*

These words clearly articulate the motley fabric that is industrialization.

The uneven development of capitalism in different countries is accompanied by an extremely uneven and simultaneous development of individual industrial sectors—each of them carrying its own unique opportunities for further development. The need to catch up with and overtake the capitalist countries requires us to examine closely the organic composition of industries in more developed countries. It immediately turns out that the production of the *means of production*, based primarily on the forces of heavy industry, constitutes the last link of industrialization. These "means of pro-

---

*Nikolai Danielson (1844–1918) published the first Russian translation of *Capital* and corresponded with Karl Marx and Friedrich Engels on capitalist economic development in and beyond Russia.

duction," in turn, gain an ever-wider scope of application. Clusters of industrialization are growing and gaining strength, turning into industrialized countries. The highest ranks of industry are beginning to acquire a distinctly international character. The prime movers of industry are becoming more and more comprehensive. The progress of industry determines the progress of transport. Transport itself is increasingly developing into a special kind of industry—an industry of transport services. Shifts in industry and transport predetermine, in turn, shifts in agriculture. In the last analysis, the entire economy becomes *a single industrial system*. The improvement of technological processes leads us in the direction of *ever more simplified and uniform technical-economic relationships*. Yet the concrete circumstances of economic development give these stages of industrialization extremely variegated forms, which makes it extremely difficult to discern the determining and decisive elements. The fate of a country's industrialization is directly dependent on several factors, and its own potentials get ground mercilessly by international relations and the timing of its own entry into capitalist circulation.

Countries entering the capitalist system have the undoubted advantage that they can use the already mature experience of their older brothers, bypassing those industrial practices stuck with the imperfection of their former technological processes. However, [international] capitalist ties are often a brake here, for the great powers of old capitalism command to different extents the will of their younger brothers.

The situation is different if an iron accelerator of events comes to the rescue at this front as well—the social revolution. Radically breaking the ties of the past, it at the same time clears the paths to the future, aligning the front of economic construction *with the front of scientific development*. This opportunity to bypass the historical stages of industrialization and *to leapfrog to its latest, critical elements constitutes our best chance to catch up and overtake the industrial countries ahead of us.*

But what are these critical, most recent elements of industrialization? A machine comes to exist when a tool is taken from the hands of the craftsman and transformed into a steel apparatus. Continuing its historical work, the machine replaces the division of labor in the human collective with a

"Energetics and Socialist Reconstruction" (1929)

complex cooperation with machines. While the working masses become a mere appendage to the machinery, a kind of aristocracy of machines and autocracy of engineering emerge. The twentieth century marks a new cycle: a new triumph of mechanization is immanent, caused by an *unprecedented concentration of productive capacities*. In the realm of machines, engines again come to the fore, which, gigantic as they are, grapple with the very elements of nature. The struggle for the automatization of handicraft skills has long been decided; on this new stage, the struggle concerns the direct control of the elements of nature. The material apparatus extends its tentacles to seize and subordinate not only heat but also the rest of nature's energy sources to human will. The transformational capacity of electricity is a great service in that regard—as Friedrich Engels has foreseen. With it ends the humiliating service of humans to the machine and the elimination of the contradictions between man and machine begins. Resolute mastery of the elements of nature cannot be reconciled with the spontaneity of living labor. Taylor* naively discovers the astonishing disproportions between the complex structure of machine cooperation and the primitive elements of living labor working under their regime. Yet the contradictions of the capitalist system do not allow its workers to create *a most effective and complex cooperation between machines and humans*, one that would be devoid of internal contradictions. Nevertheless, it is becoming more and more historically inevitable to rely on the living human being, standing above the machine: Not in the form of an individual creator of this machine—an engineer—but in the form of a conscious human collective armed with centuries of creative thought. The chances of socialism taking hold continue to grow. The element of mechanical labor, along with the remaining elements insubordinate to the will of man, are receding into the past. The aristocracy of machines and the autocracy of engineers are overcome by the rise of *socialized energetics*, unit-

---

*Fredrick Taylor (1856–1915), mechanical engineer, management consultant, and promoter of "industrial efficiency." He is widely considered the inventor of "scientific management," which seeks to improve and rationalize work processes through scientific analysis.

ing a collective of living labor devoid of internal contradictions and armed with the new energy of the elements of nature. The boundaries between the factory and higher education, between the student and the worker, are being eliminated to the extent that the economy is actually socialized. The final, critical link of industrialization, its stabilizing backbone, is *energetics—labor and its energy-equipment*—which predetermine the stages of progress towards a socialist society.

If we say that the foundation of our industrial development is the energy base, and in the concept of energetics we also include the element of human labor, we are considering possibilities of economic construction that the capitalist world is not destined to know, for the evolution of this world comes to an end *at this energetic stage* of the productive forces. These general guidelines of development do not enable us to immediately surmount the countless difficulties that stand in the way of our economic reconstruction. Yet only by sticking to these guidelines will we be able to overcome the malice of our current economic situation with the least expenditure of forces and means. Today it is bread and cast iron; tomorrow it may be fuel; the day after tomorrow perhaps textiles, and so on. The red thread of our construction is one and the same, but our current economic needs can be manifested in very different forms. To lose the red thread of construction means subjecting ourselves to all the disastrous consequences of economic opportunism. [. . .]

The concept given to us by the study of energetics and its stages constitutes the red thread of our construction: The age of steam was the age of capitalism; the age of electricity is the age of socialism.

STRUGGLE ON TWO FRONTS

The economy of pre-war Russia was typical of an industrially backward country and therefore entirely dependent on foreign capital. [. . .] Historically, the question was posed bluntly: all or nothing, social revolution or national death. The Bolshevik October [Revolution] won the battle; the task of the current generation is to fulfil this victory. This realization requires the most radical reconstruction of the entire economic system, and must not spare any relics. It should reach beyond any success that might have been achieved in the past and anticipate *the urgent future*.

"Energetics and Socialist Reconstruction" (1929)

[...]

Socialism in this sense is nothing more than a series of *great works*, transforming both the face of the Earth and the face of the workers. Its most reliable scaffold is the parallelism in the will of the working people. This parallelism is achieved insofar as a developed socialism presupposes the elimination of those internal contradictions that are characteristic of the history of labor in capitalist society. The successes of a planned economy, by which we rightly measure our progress towards a socialist system, are in the last analysis only stages in the process of transforming the conflicting, individual wills into a single powerful, purposeful stream.

[...]

Thus, we are faced with a struggle on two fronts: on the one hand, the struggle against economic opportunism, which is ready to sacrifice the most crucial perspectives of socialist construction entirely for temporary, immediate gains. On the other hand, we must fight against all sorts of adventurous attempts that neglect the necessary phases of economic reconstruction and abandon a truly dialectical understanding of economic reality.

PRODUCTION COSTS AND THE RATIONALIZATION OF PRODUCTION

Our price politics still awaits its theorist. But whatever the guidelines of such politics [...] [t]he reduction of production costs is the most urgent problem and subordinates all other problems.

[...]

Our money does not yet play the role of Marx's receipts of labor, but the evolution of our monetary-commodity relations within a planned economy can only proceed in this direction.*

---

*Krzhizhanovskii refers here to a statement by Marx in his critique of the program of the German Social-Democratic Party, the *Critique of the Gotha Programme* (1875), where he describes the relation between individual worker and society under communism: "the individual producer receives back from society—after the deductions have been made—exactly what he gives to it. What he has given to it is his individual quantum of labor. For example, the social working day consists of the sum of the individual

Gleb M. Krzhizhanovskii

In the final analysis, our price policy should strike a balance between what we may call target prices [or planned prices] and those prices that are the outcome of the natural play of market relations. The triumph of the target prices will strengthen the planning regime. Planned prices are nothing more than the expression of labor costs: prices that take into account the deciphered process of social labor. Labor expenditure is nothing more than the expenditure of living energy. Our "socially necessary time"* is nothing more than the energy costs of the economic collective that we are building. We see, thus, that with each stage we will move further away from monetary fetishism towards an energetic theory of money. Therefore, there is no longer any reason to complicate this analysis by examining the complex circumstances of our current system of circulation and price formation; we must fully focus on the decisive moments of production.

It is needless to prove that the best way to reduce labor costs per unit of production is economic planning. Suffice to say that from the very beginning of our economic development we were faced with the task of constructing not just a plan for the national economy, but a *scientific* plan for the development of the national economy. Aligning the economy with the frontier of scientific knowledge ensures maximum possible savings in the consumption of energy resources.

---

hours of work; the individual labor time of the individual producer is the part of the social working day contributed by him, his share in it. He receives a certificate from society that he has furnished such-and-such an amount of labor (after deducting his labor for the common funds); and with this certificate, he draws from the social stock of means of consumption as much as the same amount of labor cost. The same amount of labor which he has given to society in one form, he receives back in another."

*"Socially necessary labor time" is a Marxist concept that denotes the average productivity of a society: it is the amount of labor a worker with average skills, equipped with tools of an average productive potential, requires to produce a certain commodity. While Marx measured this quantity in units of time, Krzhizhanovskii here claimed that it is equal to the energy expended (a sentence that was omitted in the revised version). Krzhizhanovskii tinkered with an energy theory of labor in some of his texts, but seems never to have been fully convinced of it.

"Energetics and Socialist Reconstruction" (1929)

We see before our eyes how the economists of the bourgeois world, when analyzing losses in production and circulation, inevitably come to the conclusion that the crux of these losses lies in the fact that the scope for planning is very limited in capitalism. The large monopoly trusts undoubtedly operate according to strictly defined plans, but these plans lose their power beyond the limits of their own property.* Thus, within the capitalist system, planning occurs only as a break from private economic relations, while for us, on the contrary, private economic ties are an exception to the planned economy.

[...]

Searching in this direction [of reducing costs], we have before us the richest school of the capitalist West. But the *new* element that we must add to our [socialist] construction lies beyond the bounds of private economic progress. There is no doubt that the capacity of our branches of production, which already reveals itself to be of large absolute proportions, is a direct result of the nationalization of our industry. Nationalization avoided the lengthy [natural] process of the concentration of production capacities, which in the capitalist system takes place in the victory of the strong over the weak. In this regard, the task of the planning regime is to achieve *the technical-economic optimum in productive concentration*, in the fight against all attempts to squander social productive resources. Thus the planning regime provides only the widest possible field for typified and standardized mass production. For our socialist country it provides such prerequisites for technical and economic rationalization that in previous development stages had not been created internally by the system of its pre-revolutionary economic ties. In this sense, our organization of productive sectors has a tremendous advantage over the sectoral organizations of the West, no matter how powerful some of the Western trusts are. *Of particular importance to us are those productive advantages that we can realize by interconnecting different sectors*

---

*Krzhizhanovskii refers here to the debate on monopoly capitalism around the turn of the century. Writers like John Hobson, Rudolf Hilferding, and Lenin discussed the contemporary form of capitalism as a period in which producers sought to avoid competition by concentrating productive power across the chain of production.

Gleb M. Krzhizhanovskii

by organizing our entire industry and our entire economy on the basis of a scientific national economic plan.

The issue of mechanization by no means exhausts the entire complexity of energy transformations in industry. Finding the optimum combination of these energy transformations and carrying out the production process in such a way as to obtain the maximum useful product with a minimum of waste and losses: this is the ultimate principle of efficiency in any technological endeavor. *In other words, the most fundamental basis of efficiency in the consumption of production resources is a combination of production units that ensures the most economical overall energy balance in its sum.*

Under capitalism, the formation of the so-called horizontal trusts, spanning entire branches of industry, takes place in a long struggle and, as a rule, only partially. Even more accidental is the formation of vertical trusts, that is, interindustry links, under capitalism. It is extremely characteristic that attempts in that direction—for example by the famous Stinnes* in Germany—have failed. The deep and decisive breaking of the barriers between individual enterprises in various industries is in fact so closely associated with the concept of nationalization of industry that it cannot but scare off the capitalist world. And on the contrary, *precisely in this interindustrial connection lies the main power of the planned socialist construction of our industry.*

Thus, from this optimization of production capacities, it follows that our economic plan must first of all focus on the construction of *industrial complexes* whose energy balance will give them an enormous advantage in the competition with the capitalist world. From this point of view, the general guidelines in the struggle to reduce costs lead to the construction of *energy-industrial complexes* and to a new analytical approach to industry as a whole, *in which its energy economy is brought to the fore.*

The accounting department [*bukhgalteriia*] of our industry still awaits

---

*Hugo Stinnes (1870–1924), German industrialist from the Ruhr area, who became famous for the creation of large industrial trusts. He vertically integrated his coal company into the steel and electric industries; his Rheinisch-Westfälische Elektrizitätswerke (RWE) is still one of the four largest German energy companies.

"Energetics and Socialist Reconstruction" (1929)

reconstruction. We need numbers and statistics that with minimum recalculation can be included in the accounts of the national energy economy [*narodno-khoziaistvennoi energetiki*]. This is not to say that we should ignore the current methods of calculation, the current "production indices," but that we should quickly rise to the commanding heights of the planned economy, whose mastery is a critical weapon in the struggle between the Soviet and capitalist economies. The spontaneous rationalization of industry, [...] must be counterbalanced by a firm plan to rationalize industry that is *tightly linked to the socialist construction of the entire economy as a whole*. Until now, we have gone from the particular to the general in this matter. It is time to place the general before the particular; only then will we be able to avoid the most dangerous losses in the rate of economic growth. [...]

GENERAL ECONOMIC PLAN AND ELECTRIFICATION

We have already noted above how the development of technological processes transforms the entire national economy into a single industrial system. [...] While the practice of our "trust-making," the merging of factories and plants into trusts of local, republican and all-Union significance, is still very far from the principles of large energy-industrial complexes outlined above, our work pushes in this direction. This is most evident in the new large-scale projects that are free from the fetters of the past [such as Dnieprostroi]. [...]

Since we witnessed the electrical decomposition of matter, the lines between physics and chemistry have become increasingly blurred. Eventually, the science of matter will become part of the science of electricity. The practice of the latter has not yet matured into theory, but this will not take long. For instance, tools of mathematical analysis already enable us to overcome the gaps of the unknown and sum up the balances of the transformations of matter and energy with an accuracy far beyond our knowledge of these phenomena.

From this point of view, it is completely wrong to pit electricity's potential for mechanization against its potential associated with the development of modern chemistry. K. Steinmetz,* the greatest electrical engineer of the

---

*Charles (Karl) Steinmetz (1865–1923), German-American electrotechnical engineer

20th century, argued that the essence of electrification—its predominant principle—consists in the rationalization of the most diverse thermal processes on a broad societal scale. He expressed this idea in a famous aphorism, according to which the purpose of electrical current is to collect heat energy. [. . .]

Back in the 1880s, Fr. Engels noted with great foresight that the future role of electrification, under which the productive forces will stand in stark contrast to the capitalist way of life, rests on the fact that the electrical link constitutes the *last, final element in the cycle of energy transformations*. From here stems our view of electric "tracks" [currents] as the great "rationalizing element" of technological processes as a whole, as bearers of a new technological and social order.

If we can talk about a periodicity of the development of material culture, then we can trace these phases scientifically only by looking at the *energetic thresholds* in our struggle with the elements of nature. Marx, in his time, perfectly showed how the steam economy was the basis and stronghold of the triumph of capitalism. Engels foresaw that electrical energy brought with it social revolution, or rather, that it provided the material basis for it. Internal combustion engines are a typical representation of the transition period from capitalism to socialism. The changes they bring about before our eyes in various modes of transport and in the motorization of the rural economy, create, in turn, powerful shifts in the material basis of culture, facilitating the difficult process of agricultural socialization.

The planned electrification of the economy constitutes the crowning of these material changes towards more fully developed conditions for socialism. However, we can already foresee that a strictly centralized economic system necessary for the socialist system must in due time give way to the flourishing of free economic communes, which may come to rely on entirely

---

who worked for General Electric and was active as a technocratic socialist in the United States. He was a member of the Technical Alliance founded by Howard Scott and expressed his admiration of the Soviet electrification plan in a correspondence with Lenin.

"Energetics and Socialist Reconstruction" (1929)

new material bases. Soddy* argued that a society capable of using intra-atomic energy must reach such a level of material abundance that their entire preceding history will appear as barbarism to the people living in that era. The electric current itself, in its mysterious essence, already foreshadows the beginning of the use of intra-atomic energy even though it is still at its very rough, primitive beginnings! But it is precisely the further study of electrical phenomena that brings us ever closer to mastering the elements of intra-atomic energy. *The age of steam is the age of capitalism. The age of electricity is the age of socialism. The age of nuclear energy is the age of fully developed communism.*

---

*Fredrick Soddy (1877–1957), English chemist and Nobel Prize laureate, who together with Ernest Rutherford, discovered the elementary nature of radioactivity and studied the chemistry of radioactive substances.

7 Juan Pablo Pérez Alfonzo and the Invention of Anticolonial Democratic Oil Conservation

MICHAEL DOBSON *and* **GIULIANO GARAVINI**

IF ONE WERE TO attempt a typology of national oil governance in the twentieth and twenty-first centuries, it would vary along at least three significant axes.[1] First, in terms of governance, a spectrum would run from colonial subjugation to resource sovereignty, denoting the extent to which a national government was (and is) able, de jure and de facto, to control its hydrocarbon sector. Second, governance approaches to oil management would vary along a spectrum from "conservationist" to "laissez-faire," denoting the extent to which a nation's reserves are subject to planned and controlled development or unfettered depletion. Third, the government of the country can vary from democratic to authoritarian.[2]

There is little evidence that any of these three spectra are strongly correlated. Over the course of the twentieth century, the democratic United States saw an initially free market become dominated by a private monopoly under John D. Rockefeller's Standard Oil, witnessed efforts to establish laissez-faire competition following the company's dissolution in 1911, embraced conservationist oil management in the 1930s, and eventually returned to laissez-faire in the 1970s.[3] In this same period, anticolonial movements of the Global South, led variously by democratic, military, and authoritarian governments, consistently reclaimed natural resource sovereignty from colonial powers. In doing so, such nations have arguably acted, at times, as oil conservation actors. For example, the authoritarian government of Saudi Arabia, in consistently maintaining spare oil production capacity for geopolitical reasons, has arguably been more conservationist than many democratic oil-exporting countries, including Venezuela in the 1990s, when the policy of La Apertura (The Opening) permitted a return of foreign oil operations—and a significant increase in production.[4]

In terms of oil governance, the approach embodied by Juan Pablo Pérez Alfonzo presents something of an ideal type. Coming of age in an authoritarian Venezuela that was subaltern to Anglo-American oil capital and US political influence, Pérez Alfonzo became one of the most prominent public intellectuals and political champions of a broader Venezuelan movement for an anticolonial, democratic, conservationist oil policy. He was born in Caracas in 1903 and died in Georgetown in the United States in 1979. He lived through the beginning of the oil industry in Venezuela after World War I and witnessed its incredible expansion, a change which transformed the tropical country into the largest oil exporter in the world from 1929 until the end of the 1960s. After a brief stint studying medicine at Johns Hopkins University in Baltimore in 1922, he returned to Caracas and studied law at the Central University of Venezuela, playing a minor role in the student and popular protests of 1928 against Juan Vincente Gómez, the military ruler of Venezuela from 1908 to 1935. Gómez ruled through an ironclad partnership with foreign capital that included the "Big Three" Anglo-American oil companies: Standard Oil of New Jersey, Royal Dutch Shell, and Gulf Oil.[5]

As minister for development, a role that gave him responsibility for the Venezuelan oil sector from 1945 to 1948, and then as minister for petroleum from 1959 to 1963, Pérez Alfonzo dealt with the consequences of the transformation of the United States into a net oil importer in 1948 and with the parallel rise of the Persian Gulf as the world's foremost oil-exporting region. He witnessed the expansion of global oil consumption that rendered petroleum the world's primary energy source by the early 1960s.[6] He was worried by the "oil revolution" of 1973 when oil prices finally reached what he considered their "fair value" and when, as a consequence, a massive flood of oil rents poured into Venezuela, as well as into the rest of the world's oil exporting nations as a tsunami. Both as sponsor of the fifty-fifty profit-sharing agreement in 1948 and as a leading actor in the creation of the unique Organization of the Petroleum Exporting Countries (OPEC) in September 1960 (initiatives discussed below), Pérez Alfonzo took an active part in promoting national and international policies to deal with the transformation of Venezuela into a petro-state (a state where petroleum accounts for the vast majority of exports, and income from these exports makes up a very significant proportion of fiscal revenue). However, at the very moment of OPEC's most striking success in 1973, he became unequivocally pessimistic about the future of a country he took to defining as "mi pobre-rico Venezuela" (my poor-rich Venezuela).[7]

Pérez Alfonzo consistently studied and rationalized the meaning and significance of "petro-state," but most mainstream analysis in developed countries has displayed little interest in engaging substantively with his analysis. Only one of Pérez Alfonzo's books has been translated into English.[8] Contemporary Anglophone energy scholars are unlikely to dispute his historic significance, but they are equally unlikely to have read his work. Here, as a political theorist and historian, we are united in our belief that this neglect is unfortunate and that the technically sophisticated, democratically grounded, anticolonial-internationalist, and conservationist political thought of Pérez Alfonzo should be more widely acknowledged and studied. To achieve this, we have selected translated excerpts from his seminal and most systemic text *The Petroleum Pentagon* (1967). We provide a contextual discussion of each of the five policy angles that make up the pentagon, as well as offering further details on Pérez Alfonzo's biography.

Fully titled *El Pentágono Petrolero: La política nacionalista de defensa y conservación del petróleo* (The Petroleum Pentagon: The Nationalist Policy of Defense and Conservation of Petroleum), the book was published in 1967, following Pérez Alfonzo's retirement from active Venezuelan politics and more than two decades after he had first assumed ministerial responsibilities under the revolutionary military junta in 1945. It is a distillation of a career's worth of experience: all five angles of the unified nationalist and internationalist oil governance strategy are policies that Pérez Alfonzo had substantial experience in implementing.[9] The text is not, however, the final word on his views. As discussed below, his thinking continued to evolve in response to wider energy developments in the late 1960s and 1970s.

ANGLE ONE: FAIR SHARE

The Venezuelan oil industry of Pérez Alfonzo's youth was an extractive behemoth. In total, by 1936 Venezuela exported as much crude oil as the next seven exporting nations (the US, Peru, Iran, Romania, the Dutch East Indies, Iraq, and the Soviet Union) combined. The country was also increasingly influenced by the principles of resource nationalism that had spread through Latin America in the 1930s, under pressure from workers' movements and trade unions. After the death of Vicente Gómez in December 1935, widespread discontent against his regime exploded, and from December 1936 more than 20,000 workers started a six-month a strike over poor wages and racial discrimination in the oil-producing regions of Zulia and Falcon. From then on, oil—and obtaining a "fair share" of its value for the citizens of the country it was extracted from—became a key political issue in Venezuela.[10]

Venezuelan elites were unwilling or unable to nationalize the oil industry following the example set by Mexico in 1938: Venezuela was far too strategic for Anglo-American oil companies to accept the loss of direct control over Maracaibo, and at the same time its government had become overly dependent on oil revenues to risk economic boycott. But in the post-Gomez era military rulers and technocrats such as Manuel Egaña designed new laws that radically altered the subaltern relationship between the Venezuelan state and Anglo-American oil companies that had persisted under Gomez's rule, in which these companies obtained "concession contacts" over vast tracts of the country. (For more on these lopsided contracts and efforts to supersede them, see "Angle Four: No More Concessions" below). A milestone in this process was the approval of a new oil law in 1943 by the government headed by General Isaías Medina Angarita. This legislation allowed for an increase of up to one-sixth in the royalties on oil production (16.67 percent, a higher level than in the United States), canceled all tariff exemptions, and reaffirmed Venezuela's complete sovereignty over fiscal policy.[11] Even though the law had been negotiated with the Big Three oil companies, it also benefited from the US's strategic interest in appeasing its crucial Latin American ally while the Second World War raged in Europe and the Pacific, and Medina Angarita's affirmation of fiscal sovereignty in particular was of fundamental importance for subsequent governments.

By the end of the 1930s, Pérez Alfonzo had become a renowned university professor in law and had moved increasingly close to the politician Rómulo

Betancourt, the strongest political figure in Venezuela's democratic movement. After Betancourt founded Acción Democrática (AD) in 1941, Pérez Alfonzo became the party's voice on the crucial issue of oil policy. During the 1943 oil law vote, while expressing appreciation for its innovations, he voted against it (the famous *voto salvado*), objecting that the approach to taxation would not ensure an adequate share of the oil rent for the state if crude prices increased beyond a certain level and also that the law condoned all the past illegal actions of these oil companies.[12] The speech earned Pérez Alfonzo national attention, striking a chord across multiple constituencies, including the army.[13]

In October 1945, a coup removed Medina Angarita from power and the new military establishment appointed a revolutionary junta presided over by Betancourt. Pérez Alfonzo became minister for development (with direct responsibility for the oil sector) and was later confirmed in this position in 1948, after Venezuela's first democratic elections led to the presidency of the famed novelist Rómulo Gallegos in February 1948. For three years Pérez Alfonzo held a key decision-making position with regard to the oil policy of his country. In this position he facilitated two policy decisions with a profound influence on Venezuela's attitude towards oil extractivism and on the global oil industry in general. One of these—the policy of "no more concessions"—is the fourth angle of the pentagon, discussed below.

The other key decision, widely known as the fifty-fifty profit-sharing model, was adopted in November 1948. In simple terms, the new tax regime established that, in order to ensure a "fair share," net profits deriving from the sale of Venezuelan crude had to be evenly split between the state and the concessionary companies. According to Pérez Alfonzo, the oil industry's reliance on Venezuelan natural resources meant it had to be considered a "public service," because its profits derived not only from capital investment but also from the appropriation of an international land rent that foreign capital had not generated. Accordingly, it decreed that the industry's profit rate should be limited to no more than 15 percent per annum. Pérez Alfonzo identified the interests of Venezuela as those of a sovereign landlord collecting a rent (consisting in the "reasonable participation of the nation as the owner of the oil fields exploited by the industry") that could eventually be redistributed for development purposes, such as improvements of wages both for workers and public employees, tax cuts on small and medium industry, improved infrastructure, and the promotion of Venezuelan industry. The fifty-fifty model established a partnership between international oil companies and the Venezuelan state that aimed at limiting the profits of foreign investors in their dual role as capitalists and managers of a public service. (Ultimately, however, it also redounded to these companies' benefit by setting a roof on the taxes that could be imposed on the Big Three.) This model spread from Venezuela to most other oil-exporting countries by the early 1950s.

ANGLE TWO: COORDINATING COMMISSION FOR THE CONSERVATION AND COMMERCE OF HYDROCARBONS (CCCCH)

At the end of 1948, a military coup dismissed Venezuela's first democratically elected government. Pérez Alfonzo and other democratic leaders were forced into a decade of exile. Pérez Alfonzo lived for a time in both Mexico and the United States. In the former, he was able to examine the functioning of a fully nationalized oil industry. In the latter, he could observe firsthand (as had Saudi technocrat Abdullah Tariki a few years earlier) the world's largest oil industry at the time and the conservationist philosophy and institutions that governed it. The most notable of these was the Texas Railroad Commission (TRC), the elected three-member commission that controlled the great balance wheel of Texan oil production from the 1930s, adjusting the state's huge output on a monthly basis to ensure that overall US oil supply matched demand, thus keeping prices steady. This practice of production rationing, or "prorationing," was an American innovation, albeit one grounded in a conservationist ideology with deep intellectual and political roots, stretching back to Progressive Era concerns about the exhaustion of resources such as coal and lumber.[14]

In 1958, a Venezuelan military government was once again toppled by forces that promised and delivered free and fair elections. Rómulo Betancourt was elected president and promptly returned Pérez Alfonzo to his ministerial post. Even before the elections, the provisional government, emboldened by violent demonstrations against the visit to Caracas of US vice president Richard Nixon (and pressured by significant budget constraints), issued a decree raising taxes for the oil companies, increasing the government's take from the agreed 50:50 ratio to 64:36—a fundamental reinterpretation of the appropriate "fair share" focused on recapturing excess profits that concessionaires were making due to a massive increase in petroleum production between 1950 to 1957 and to the favorable prices.

Pérez Alfonzo created the CCCCH, the second angle in the pentagon, in 1958, explicitly modelled on Texas's TRC. It aimed at controlling the rate of production and (thus) international prices of Venezuelan oil. Matching supply to demand requires cooperation by all major producers in the relevant market, however. Within the US, this occurred from 1935 through the Interstate Compact to Conserve Oil and Gas, which allowed state authorities including the TRC to informally coordinate production on a national basis, although no binding quotas were set. For the overwhelmingly export-oriented industry in Venezuela, however, an equivalent level of cooperation required, as Pérez Alfonzo noted, "concurrent action of the other countries involved in international trade." It is for this reason that he considered the CCCCH's "main success was acting as a stimulus for the creation of OPEC," as discussed further below in "Angle Five: OPEC." In enabling this transcultural migration of petroleum conservation ideology and practice, first to the domestic Venezuelan context and subsequently to the international level, Pérez Alfonzo made a significant contribution to energy history.

ANGLE THREE: VENEZUELAN OIL COMPANY (CORPORACIÓN VENEZOLANA DE PETRÓLEO [CVP])

For Pérez Alfonzo, the creation of Venezuela's national oil company had an important pedagogical, as well as economic, function. The country's state oil company was initially tiny compared to the Big Three it operated alongside, but Pérez Alfonzo realized that managing the CVP would require Venezuelans to become familiar with every aspect of the oil business, including operating costs, technological innovation, marketing and refining—a gamut of tasks that provided an opportunity to overcome their discrepancy in technical expertise and to compete with the international companies. The CVP would eventually be able to engage these international companies through "service contracts," however, allowing the country to continue benefiting from foreign expertise and investment, without perpetuating a system of concession contracts that Pérez Alfonzo viewed as fundamentally exploitative and unacceptable.

As we have mentioned, Venezuela was not the first Latin American country to create a national oil company (importantly Petróleos Mexicanos, or

PEMEX, had been created in 1938, and the Bolsheviks had nationalized oil soon after the 1917 revolution), and state oil companies had also been created earlier in other industrialized countries such as Italy (ENI in 1953). But it is still important that the largest oil exporting country in the world aimed at directly operating the industry in the medium term through a national company.

ANGLE FOUR: NO MORE CONCESSIONS

The principle of granting no new concessions—a type of contract dating back to the days of the Dutch East India Company, which allowed international companies to own and exploit resources within a given area of land, typically for decades or until the complete exhaustion of reservoirs and mines—was applied unfailingly throughout both of Pérez Alfonzo's tenures in government.[15] During his intervening exile, he wrote to President Pérez Jiménez urging him to adhere to the same policy.[16] His efforts were unsuccessful, and new concessions were awarded in 1956 and 1957.

An end to granting concessions did not mean an end to foreign investment in the oil sector—at least not for the Pérez Alfonzo. It simply meant bringing production under the control of the state and allowing exploitation of new reserves by foreign companies only through service contracts. During his second tenure as minister for petroleum, however, he was increasingly critical of the influence that foreign investors were having in his country and became further radicalized after he left office. Following a failed effort at fiscal reform in 1966 (aimed at recouping the, in his view, unfair profits of the oil companies), Pérez Alfonzo definitively lost hope in the possibility of cooperating with the Big Three.

Influenced by the widespread criticism in Latin America (and beyond) against imperialism and international capitalism, Pérez Alfonzo's arguments against dependency and foreign capital were systematized in a book published in 1971 with a double title: *Petróleo y Dependencia. Petróleo y Desarrollo Económico* (Petroleum and Dependence. Petroleum and Economic Development).[17] Foreign investors, he wrote, represented a country within the country, protected by massive US influence (the US Venezuelan embassy alone hosted 250 employees) and by deep connections with the conservative creole elite. Venezuelans were still forced to pay the *diezmo colonial* (the "diezmo" was a tax collected by the Spanish king in colonial times) to the

US, since a yearly average 10 percent of Venezuelan GDP (the value of the profits of foreign investors) was transferred abroad. "I believe," he would comment, "that for Venezuela, more than for Canada, there is only a dual option 'independence or colony.'"[18] The oil companies fought assiduously against efforts to limit their profits, while pressing to increase production, seemingly indifferent to the low international price of crude oil. According to Pérez Alfonzo, this stance was also contrary to the interests of US citizens, since cheap oil kept global demand for energy continuously on the rise, and "massive consumption means at the same time massive pollution."[19]

Foreign capital also forced Caracas to become increasingly dependent on oil revenues. The rate of growth in oil income was consistently higher than the rate of growth of the non-oil economy: by the 1960s, for Pérez Alfonzo, it had clearly become impossible to transform oil income into productive investments and capital formation. The liquidation of Venezuelan petroleum, a nonrenewable resource, mostly translated into an increase in private wealth and ever-increasing and irrational (*despilfarro*) state expenditures, rather than serving the public good.

This ultimately led Pérez Alfonzo to propose that a permanent ceiling be placed on oil production to limit the amount of revenue this generated—and then, famously, to declare that oil was *el excrement del diablo* (the devil's excrement). However, this shift of view on the energy resource were less a reversal than an evolution, as long-standing preoccupations of his thought, which could already be seen in the excerpts from this section of the pentagon, came to dominate his thinking. In an interview in 1975, Pérez Alfonzo concerns about the wastefulness of oil wealth took center stage, leading him to make an impassioned plea:

> I have consciously lived the entire history of oil in my country—frequently participating directly in the matter—and for this reason my experiences and observations force me to insist on the need to put a stop to this deforming activity.[20]

ANGLE FIVE: OPEC

While the peculiar natural resource endowments of Venezuela and the Persian Gulf provided a basis for cooperation, OPEC was by no means geologically foreordained. Political solidarity was built, not found, and the birth of the organization was inseparable from both the conservationist ideology that inspired it and from the spirit of the ongoing anticolonial struggles alongside which it emerged.

Although Pérez Alfonzo had met with the Iranian ambassador in Washington, D.C., in 1947, it was not until 1949, after the fall of Venezuela's first democratic government, that a group of Venezuelan technocrats first travelled to the Middle East, finding a particularly receptive audience in Tehran.[21] Iran's subsequent nationalization of the industry and Prime Minister Mohammad Mosaddegh's struggle for Iranian oil sovereignty at the UN General Assembly and International Court of Justice signaled a shift of momentum towards the Middle East. As the international oil companies responded with a boycott that saw Iranian production plummet (while Iraqi, Kuwaiti, and Saudi production made up the difference), the British convinced newly elected US President Eisenhower to support Mossadegh's overthrow in 1953, and momentum shifted again. This time the struggle moved to Cairo, where Gamal Abdel Nasser, emboldened by his Suez triumph, sought to unite Arab oil producers, and to Saudi Arabia, where an emerging technocratic movement headed by Abdullah Tariki sought to contest the power of the mighty US firm Arabian American Oil Company (ARAMCO).[22] Upon Pérez Alfonzo's return to government in 1959, the stage was set for a producer's alliance to begin.[23]

Invited as an observer to the first Arab Petroleum Congress in 1959 in Cairo, then capital of the United Arab Republic, Pérez Alfonzo worked with Tariki to corral representatives from Iran, Kuwait, and the UAR into signing a "gentlemen's agreement" that recommended the creation of an intergovernmental Oil Consultation Commission and recorded agreement on five key principles, including the establishment of national oil companies and the creation of "organisms for the conservation, production and exploitation of petroleum," as well as the improvement of "oil producing countries' participation on a reasonable and equitable basis," with a new target of a 60:40 profit split with the companies. It was this document that provided the blueprint for the creation of OPEC seventeen months later.[24]

At a time when the market was exerting increasing downward pressure on the price of oil and when the US had introduced mandatory quotas to protect domestic producers, Pérez Alfonzo and Tariki considered US independent producers a potential ally in the effort to defend the oil price. In April 1960 they worked on a paper titled "The Concept of International Petroleum Proration" that Tariki would deliver at the annual meeting of the Organization of Texas Independent Producers on May 2.[25]

The speech claimed that their objective was to stabilize the oil market and promote the conservation of a nonrenewable natural resource for the benefit of their countries and for future generations of consumers. It reassured the audience that nobody had to fear unlimited price increases, since there were enormous shale reserves in the United States and Canada and "it would be the height of economic folly to either encourage or even force, through the economic abuse of proration, the development of resources which could conceivably supplant the present sources."[26] A global proration formula for oil should not raise any eyebrows since it was "a form of economic unity that has all the justification of the European economic blocs or of the International Coffee Convention." On the contrary, the speech underlined that "what makes proration even more important is the fact that it is protective of a wasting resource of tremendous importance to the world."[27] Following his delivery, Tariki flew to Caracas, where he presented the plan together with Pérez Alfonzo and Mohamad Salman, the Iraqi engineer who directed the petroleum bureau of the Arab League. In making these connections, Pérez Alfonzo reaffirmed that economic cooperation between oil-exporting countries was crucial to the effective conservation of petroleum.[28]

The decision of Standard Oil of New Jersey (today Exxon) to unilaterally reduce the "posted prices" for oil in April 1960 triggered the creation of OPEC in Baghdad that September. This is not the place to recount the details of the negotiations that led to the Baghdad conference, and we do not possess a reliable record of the conversations in Baghdad, since the collected minutes of the OPEC Conferences do not include the first meeting (possibly because OPEC's bureaucracy had not been established yet).[29] We do know from the resolutions approved at the end of the founding conference that all the participants, well aware of the key role played by oil as a primary energy source and of their dependence on the oil rent, aimed at stabilizing the market and restoring the posted prices to their position before the price cuts of 1960. What is less well remembered is that global prorationing was also a key objective in the whole cooperation effort: "Members shall study and formulate a system to ensure the stabilization of prices by, among other means, the regulation of production," OPEC's founding treaty records, albeit this objective was to be achieved while safeguarding the interest of the "consuming nations," providing "efficient, economic and regular supply," and assuring a "fair return to capital."[30]

Pérez Alfonzo died from pancreatic cancer on 3 September 1979.[31] An article in *The Washington Post* two weeks later styled him as "The Man Who Invented OPEC—To Conserve Energy," highlighting his view that the gas-guzzling automobile was a "cosmic curse" and noting that while many focused on OPEC as a means to increase oil prices, he saw it as "a way to lower the use of energy."[32] For key protagonists in the creation of OPEC, and especially for Pérez Alfonzo, prorationing and resource conservation were at the very heart of the organization's intended operation. This derived partly from the fact that in Venezuela the oil companies were taxed on the market price of crude oil (as opposed to the other OPEC founders, where taxes were paid on the basis of "posted prices"). Caracas was willing to accept a decline in oil production growth, so long as this was combined with an increase in market prices or at least with a mechanism to prevent their decline. For Venezuela, then the largest oil exporter in the world (but not for much longer), prices came before production.

Prices and the guarantee of income for the national budget were only one aspect of the Venezuelan strategy. The other component, embodied in the "no more concessions" policy, was the idea that oil was a nonrenewable natural resource which had to be preserved for future generations. For this reason, while foreign investors were still to be welcomed (once they accepted the role of public service providers), there had to be clear limits to production increases, as much as there had to be limits to their profit margins.

This aspect of Pérez Alfonzo's politics finds an (unacknowledged) resonance in contemporary "supply side climate policy," which seeks a managed decline of the fossil fuel industry in line with the imperative of mitigating dangerous climate change. In May 2021, the International Energy Agency (IEA) released an analysis of the changes to the global energy landscape required to achieve net zero greenhouse gas emissions by 2050. The report stated that "the trajectory of oil demand in the [net zero emission scenario] means that no exploration for new resources is required and, other than fields already approved for development, no new oil fields are necessary," since production would have shrunk from 100 million barrels of oil per day in 2019 to 24 million per day by 2050.[33] Asked for his opinion, Saudi oil minister Prince Abdulaziz bin Salman described it as "a sequel to the film *La La Land*," an unrealistic fantasy.[34] This attitude was in line with other leaders of OPEC, which suggested managed decline of global oil production, to meet the imperatives of climate change, is unlikely to be endorsed by the organization.

In this context it is useful to remember that prorationing, a supply-side energy policy based on setting oil production quotas in order to prevent "economic waste" and to promote resource conservation, has been a key feature of the most important oil producer and consumer in the world, the United States. This was particularly true from the 1930s to the early 1970s when the TRC, by controlling production in the most important oil-producing state within the US, was a reference point for global crude oil prices. The TRC, as we have seen, was a model for Pérez Alfonzo, who deserves wide recognition not only as the authoritative voice of the largest oil exporter in the world for forty years and as a "founder of OPEC" but also as a technocrat who in the 1960s increasingly identified with the "environmentalist" movement and its priorities. His thought and praxis endure today as a potent source of internal critique of both OPEC and other nations' dependency not just on oil but also on the critical minerals that will be needed in the "energy transition" away from fossil fuels.[35]

JUAN PABLO PÉREZ ALFONZO
The Petroleum Pentagon (1967)*

The nationalist oil policy can be set in a framework within what has been called the "Pentagon of Action" with five key angles. One of these is the Organization of Petroleum Exporting Countries (OPEC), which extends its lines beyond the sphere of national jurisdiction. The other four angles of the structure of the "Pentagon of Action" remain completely under Venezuelan sovereignty and are formed by the directive principles of: 1. Fair Share;

---

*Juan Pablo Pérez Alfonzo, *The Petroleum Pentagon*, trans. Karen Sturges-Vera (Vienna: OPEC, 2003), extracts from pp. 22–25, 41–43, 59, 73–87, 89–91. Originally published as Juan Pablo Pérez Alfonzo, *El Pentágono Petrolero. Política nacionalista de defensa y conservación del petróleo persigue liberar al país de la excesiva dependencia de un solo recurso no renovable* (Caracas: Ediciones Revista Política, 1967). Permission for reprinting granted by OPEC.

The Petroleum Pentagon (1967)

2. Coordination Commission for the Conservation and Commerce of Hydrocarbons; 3. Venezuelan Petroleum Corporation; and 4. No More Concessions. Naturally, the logical order of the five fundamental principles of the defense and conservation of petroleum begin[s] with the four that fall under the complete jurisdiction of the country, in the order put forth above. They are complemented by the principle of international relations, based on the community of interests that served to form the already known and powerful Organization of Petroleum Exporting Countries.

[...]

ONE: FAIR SHARE

In petroleum policy, as in the policy for the use of any natural resource that is collectively owned, the question of obtaining a fair share in the exploitation of the collective wealth is the first question that comes forward for the administration to which the State has entrusted it. But for Venezuela, obtaining a fair share of its oil, coveted by the powerful and adept foreign interests, has been a fight that is still unfinished. While the national sovereignty was usurped by the dictatorships of the moment, foreign interests moved at their discretion and convenience in the face of the negligence and fear of the usurpers. For their action, the foreign interests counted on their own powerful forces and their well-tried ability in the enjoyment of the impotence of others. [. . .] This interest of the governments of the large investment countries (which at the same time are great consumers of oil) increases as they become more dependent on the international trade of this source of energy. Thus it is not surprising that a continual battle is required to maintain a fair share, given the changing dynamic of the conditions of the oil industry. All of this is directly related to the defense of prices, the core and basis of this share. At the same time, the main objective of the buyers is to operate at the lowest prices, always forgetting what in words they offer to the poor countries as a contribution to their development.

[. . .]

In its first period (18 October 1945–24 November 1948), the Democratic Government hardly had time to begin to define a fair share and to formulate its basic principles. For the first time they began to investigate the accounts of the companies and to analyze the state of profits and losses resulting from their accounting. Thus the bases were given for determining the limits of what could be the fair share, in function of the net profit justified by the capital invested to produce the profits from oil exploitation. As an equitable application of this principle, Decree No. 112 of December 31, 1945, was announced. Circumstantially, the conditions of the moment resulted in that the raising of the share to a level at least equal to the companies' net profit also lowered that net profit to a level acceptable in relation to the capital. At the same time, the share increased considerably. That was what was then called the 50–50 share. [. . .]

From this same time, the Democratic Government applied the economic doctrine of public service and the consequent fair profit to the existing system of oil concessions, which determined the fair share. In addition, the Government initiated the defense of prices, which are the bases for providing the profits and the share. For the first time in the history of the world oil concession system, a Government decided to receive in kind the royalties that the system attributed to it, to offer them for public sale on the international markets. The predominant situation in the oil markets after the end of the Second World War was a demand that increased faster than the possible production and transport. In all, the international consumers tried to retain the level of oil prices frozen during the war, regardless of the market situation and the soaring rise of prices of manufactured goods. Since then, the story—now and always—is of the success of the strong to sell dear to the weak while buying from them as cheaply as possible, without any compensation of any kind. With the tendering of the Venezuelan royalties, the siege of frozen prices broke, and so earnings increased in 1948 and 1949 to reasonable levels, improving the country's participation in the exploitation of its fundamental collective wealth.

[. . .]

The Petroleum Pentagon (1967)

TWO: COORDINATING COMMISSION FOR THE CONSERVATION AND COMMERCE OF HYDROCARBONS

The Coordinating Commission for the Conservation and Commerce of Hydrocarbons forms the second angle of the "Pentagon of Action" of the petroleum policy. It is the instrument to defend prices to avoid the economic waste of oil, which is depletable and nonrenewable. Since 1945, the Democratic Government has recognized the need to take a more direct part in the economy of the oil industry that administers the granted concessions. The review and consolidation of the concessionaires' accounts were begun as a basis for determining the fair share that should be obtained from the oil. The fair share required intervening in the industry's economy to see how and how far the profits obtained should be divided. This logically leads to the question of defending prices, which is the economic core of all industrial activity.

From that time, the defense of prices required that the State act vigilantly, and thus it was decided to invite tenders for the oil royalties, to determine precisely to what point the market would raise prices above the arbitrary levels fixed in the published quotations. The success of this action was evident, and it engraved on the Venezuelan mind the need for the State to take an active role in the economy of oil concessionaires. For this reason, upon assuming power, the Constitutional Government was compelled to create the Coordinating Commission as a permanent instrument of action in the defense of prices, the basis of the industry's economy and consequently the State's share. [. . .]

The prevailing conditions, showing an excess production capacity available in all world production centers, required coordinating production with possible demand. This demand was constantly rising but it was inelastic to the effects of price changes. The Coordinating Commission has filled such a role in Venezuela, and it is expected that other producing centers will follow this example, which the conservation of oil resources requires. [. . .]

The constant need to coordinate production with demand, first attended to in the United States and Canada and later in Venezuela, is a need imposed by the nature of the oil trade which soon, as already stated, will have to be extended to the other oil producers, especially to those involved in the international trade of the product. As a first step towards this universalization of

Juan Pablo Pérez Alfonzo

coordinating production to adjust to demand, in 1960 the Organization of Petroleum Exporting Countries was created, which will be discussed later. The earlier creation and operation of the Coordinating Commission in Venezuela was a good stimulus for the creation of that Organization.

As is easy to understand, the success of the Coordinating Commission was seen to be limited since Venezuelan oil production has been directed mainly towards international trade. The effectiveness of the action outside of the national arena partly depends on the concurrent action of the other countries involved in international trade. In that sense, the Coordinating Commission's main success was acting as a stimulus for the creation of OPEC, which in turn is laying the foundation that in the near future will allow the joint action of all countries interested in the defense of prices.

[...]

THREE: VENEZUELAN PETROLEUM CORPORATION

The Venezuelan Petroleum Corporation is another angle of the policy of defense and conservation of petroleum. After so many years having foreign investors initiate and develop this national richness, the desire to organize and put in place a national enterprise managed by Venezuelans, for Venezuelans was becoming urgent. Since 1948 this ambition was close to being realized. Its postponement during the dictatorship was only one of the evils of that usurpation of national sovereignty. The CVP was created in April of 1960 and the following year it was functioning.

[...]

FOUR: THE PRINCIPLE OF "NO MORE CONCESSIONS"

The principle of no more concessions, like that of fair share, was formulated in 1946, when for the first time the Government acted as the authorized representative of the Venezuelan people.

[...]

The skillful and interested propaganda that for so long has been conditioning Venezuelans to act as if the current wealth were inexhaustible, must be counteracted with will, carefully analyzing the entire situation. Because the country is blinded by its wealth, it wastes it without worry, making it

The Petroleum Pentagon (1967)

easier for the businessman to exploit the business to liquidate it in the shortest term possible. If time generally is gold, in oil it is even more . . . for the concessionaire. Nevertheless, for the owner of the oil the time is gold in the contrary sense, in many ways. Time that passes is time that earns, because the oil, an exhaustible resource, becomes more and more expensive to find and produce, while more and more the necessities of progress and the aspirations for well-being of the people demand it in greater quantities.

To the greater future value is added the possibility that with time a more reasonable division of profits is made. Furthermore, the certain future value is not compensated with the current useful value, because generally the oil exporting countries have exceeded their real capacity for employment of capital. Precisely: the countries have exceeded their capacity as a consequence of the income they receive and that, even though most frequently it would not have been possible to reach a reasonable level, the income is so important and grows so rapidly that it is difficult to use it in the most advisable way. [. . .]

The substitution of the system has been conceived and planned since the same moment in which the termination of the system of concessions was proposed. This substitution is underway with the Venezuelan Petroleum Corporation, and the complementation of the new system with the service contracts should be done at any moment. [. . .]

FIVE: ORGANIZATION OF PETROLEUM EXPORTING COUNTRIES (OPEC)

The principle of the union of all oil exporting countries comes from the community of interests and the necessary self-defense against the powerful foreign forces, constituted by the international oil industry that manages the national riches. The urgency of rapprochement between Venezuela and the Middle East was clearly seen once the World War was over, the US oil industry entered into decline, and the unexpected post-war demand accelerated the placement in production of the enormous oil reserves in the Middle East.

In 1947 the great oil country, the United States, crossed the line of being an exporting country and became a net importer of oil. In the same period, the Venezuelan Government took a more active position vis-à-vis its oil wealth exploited by foreign companies. The Government began action on the question of prices, that were expected to be kept frozen due to the pressures of the consumers market. On the other hand, the Government also was moved in the sense of imposing a fair share on the profit obtained by foreign investors. After that, the threat of Middle East oil, with its minimum costs and incalculable reserves, was a constant element of pressure to try to return to the previous situation. But also after that, Venezuela was aware that the community of interests would rise in solidarity at the head of the defense against the opposing interests.

Commenting on a news article in *Oil and Gas Journal* about the increase in Middle East production, the October 19, 1948, issue of the Information Bulletin of the Ministry of Development stated: "Venezuela really is pleased that the other countries can enter in to share with her in the very serious responsibility of supplying the world with its growing need for liquid fuel. . . ." After that, the correct position was assumed of not launching into an unlimited production war. On the contrary, the need to limit Venezuela's participation in the international markets to a reasonable concurrence with other sources of supply was recognized.

Soon it was seen that the Middle East also would come out in defense of its oil and that the great distances separating the Caribbean Sea and the Persian Gulf would not impede the coincidence of interests. After Venezuela raised its share by means of the tax system called "50–50 Arrangement," the Government of Iran solicited an equivalent share from the Anglo-Persian Oil Company, predecessor of British Petroleum. In 1949 the Anglo-Iranian [Oil Company] tried to reach an arrangement that would keep the Persian fair share secret, but the Parliament rejected it. Thus arose the process that finally led to nationalization, when Mohamed Mossadeq assumed power in 1951. The question planted then as a consequence of Iran's having tried to obtain conditions similar to those of Venezuela would have followed a different course if Venezuela and other exporting countries would have had responsible governments in those years, aware of the common interest that was in play in the case of Iran. Unfortunately, it was not so and Mossadeq was defeated by isolation and the belt of economic strangulation that oppressed Iran, together with the mistakes of exaggerated nationalism. Even

The Petroleum Pentagon (1967)

so, the Venezuelan action, followed by Iran's attempt, produced favorable consequences for the other exporting countries. [...]

By way of all of these events, the peoples of the oil exporting countries were becoming aware of the need to unite their efforts for the defense of common interests. It would be a question of waiting for the opportunity of responsible governments that would give a boost to the process of rapprochement. [...]

8 Privatizing a Colonial Electricity Undertaking

F. W. Dove's "What People Think of Our Electric Light"

DAMILOLA ADEBAYO

IN MARCH 1909, Frederick William Dove, barrister, businessman, and leading member of the early twentieth-century educated West African elite, made a case for privatizing the first power station in Lagos, a British colonial territory in today's southern Nigeria. In the newspaper article presented here, which Dove described as a product of "a careful study of the Electrical question at Lagos and in Europe during the greater part of the last twelve months," he proposed that the electrification of Lagos should be put "in the hands of a Private Company." He argued, among other things, that the colonial government was indifferent to the abysmal access to electricity in the city at the time because it was not responsible to "shareholders." Dove's article is the first documented call by an African to privatize a state-owned energy enterprise in colonial Africa. Moreover, his proposition was also unusually prescient because only a few African countries had electricity supply in 1909 and subsequent uncertainties about the political economy of electricity have affected contemporary electricity supply in Nigeria.

This chapter examines Dove's statements on electricity management in early twentieth-century Lagos in terms of their implications for political economy. The focus is his positionality as an upper-class African and the implication of his privatization thesis for ordinary residents. Here, we see an

African "nationalist" in favor of privatization when everyone else in Lagos was demanding additional government investment in utilities.[1] I argue that Dove's privatization proposal *inadvertently* presented him as a figure of authority who did *not* regard electricity as a public good. His proposal, if realized, would have excluded more Africans from accessing electricity because private firms would have likely raised energy costs to recoup their investments and maximize profits. Nevertheless, his critique of public management is germane to later debates about electricity privatization in twenty-first-century Nigeria. In fact, Dove might be considered a Cassandra who foresaw the challenges of public management, particularly in terms of the risks of inefficient service delivery and waste in government spending, but his privatization "prophecy" came a century too early.

The rest of the paper is structured as follows. The second section summarizes the socioeconomic context that inspired Dove's article. The third section assesses the merits of Dove's arguments. It provides insight into the challenges of colonial-government-managed electrification in early twentieth-century Lagos and explains why Dove believed private companies were better positioned to run the nascent electricity industry. The fourth section critiques Dove's reasoning by analyzing the implications of privatization for the average African and explores his biography to query his public interest claims. The fifth reviews the significance of Dove's proposal for contemporary debates about the privatization of the Nigerian electricity supply industry. The paper concludes by situating Dove's arguments within the understudied history of early twentieth-century Nigerian economic thought.

SOCIOECONOMIC CONTEXT

By the time Dove began writing his article in 1909, Lagos had been electrified for eleven years. The first power plant became operational in early 1898.[2] The generating machinery was steam-powered, with water drawn from the Lagos lagoon and coal as fuel. It consisted of two British-designed Paxman Locomotive–type boilers with two compound double-acting engines, driven by means of a bolt, and a 30-kilowatt single-phase 80-cycle alternator, all of which generated electricity at 1,000 volts. This electricity was distributed through switchboards to overhead circuits. A circuit supplied 20 incandescent streetlights, each equivalent to the light of 50 candles. The space between each ranged from 100 yards (in the predominantly European-populated streets) to 140 yards in other neighborhoods. Between 1898 and 1909, this capacity was increased to 120 kW to meet growing demand. As of August 1909, the power station consumed an average of 77 tons of coal monthly.[3]

The electrification of Lagos was state-led.[4] The government introduced electricity primarily for lighting official residences, streets, and homes, with the hope that it would spur future industrial activities. Before 1898, Lagos had been mainly illuminated by lamps fueled by palm oil.[5] However, electric light was regarded as a "cleaner" and "brighter" technology that befitted a modern city like Lagos.[6] The city's residents, in particular, believed that the electric light would bolster Lagos's status, which it had earned during the 1880s, as "the most progressive of H.M.'s possessions in West Africa."[7]

Lagos was a modern colonial city by the 1880s due to specific geographical advantages, which, in turn, helped shape the outlook of its inhabitants. The city had evolved in the fourteenth century from a small island (now Lagos Island) of less than two square miles.[8] It is situated in the Bight of Benin, the curved coast of West Africa, an area notorious for supplying millions of African captives during the trans-Atlantic slave trade.[9] The British bombarded Lagos in 1851, allegedly to oust the island's king, Oba Kosoko.[10] The king had defied the Act for the Abolition of the Slave Trade, which Britain's Parliament had passed in 1807. Lagos officially became a British colony in 1861.

In subsequent decades, Lagos became a principal center of colonial West African maritime commerce. The city retained all revenues from customs duties levied on commodities passing through until 1914, when it was incorporated into colonial Nigeria. Lagos was a wealthy city with, for instance, treasury surpluses of £34,135 in 1894, an investment of £48,000 in London also in 1894, and surpluses in previous years. This money helped fund ambitious public works in transportation, communication, and health.[11] Lagos remained Nigeria's economic and political capital throughout the colonial period.[12]

The socioeconomic opportunities available in the city led to a continuous wave of immigration. Lagos had an estimated population of 28,513 in 1871; 37,452 in 1881; 41,847 in 1901; and 73,000 in 1911.[13] In 1963, it became the first West African city with a population that exceeded one million.[14] Lagos expanded beyond the island and, by 1911, had incorporated neighboring mainland villages into a municipal area of roughly 18 square miles.[15] By the late colonial period (circa 1950), its landmass had increased to 27 square miles.[16]

Lagos has since continued to expand, primarily through land reclamation, to become a megacity of 450 square miles.[17] According to one government estimate, today this expanding metropolis has over 21 million residents.[18]

Trappings of colonial modernity, including the telegraph, harbor works, railways, schools, bridges, and hospitals, were introduced from the 1860s onward.[19] More importantly, late nineteenth-century Lagos could be said to be modern because of its African inhabitants' lifestyle.[20] During the 1900s, Lagos's residents comprised original Yoruba settlers, "liberated Africans" (Africans rescued from slave ships by the British navy after the 1807 abolition), and emancipated African returnees. The latter came to Lagos from Sierra Leone, Liberia, Brazil, Jamaica, and the United States. A general feature of Lagos's African population was their conspicuous consumption of European goods, fashion, and ideas—or what Bronwen Everill calls "material cosmopolitanism."[21] They also had access to primary and secondary schools built by Christian missionaries from the 1840s onwards.[22] Some obtained degrees from higher institutions abroad, particularly at Fourah Bay College, Sierra Leone, and British universities.[23] Lastly, some Africans, particularly Afrodescendant settlers, possessed advanced European craft skills (as bakers, tailors, and masons, among others).[24]

Consequently, African middle and upper classes "dominated the legal profession in Lagos, as well as being the backbone of the churches and the colonial service."[25] Unlike settler colonies in Eastern and Southern Africa, by the late nineteenth century, Lagos "natives" were shaping colonial policies as legislators, were prominent in racially hierarchical occupations such as medicine, and were leading cash-crop farmers and agents of European trading firms.[26] Fredrick William Dove was part of this sociopolitically conscious generation. Other leading nationalists of the time included Orisadipe Obasa, Adeyemo Alakija, John K. Randle, Kitoyi Ajasa, Richard Akinwande Savage, and Sapara Williams.[27]

Dove and his contemporaries felt entitled to electricity. They had been learning about electricity principles since the 1860s and anticipating electric lighting since 1889, when political debates on how it should be funded began.[28] Funding for electricity came from indirect taxes on alcoholic spirits levied on all Lagosians. However, when supply commenced, Dove argued the government prioritized official residences, "mercantile firms and a few native gentlemen who could afford to pay" despite the initial rhetoric of "public interest." He was not alone: African demands for electricity were rife during the early 1900s. Newspaper publications were their principal medium of expression and activism.[29] In Lagos, journalists argued that electricity rates were prohibitively high, and they blamed the Public Works Department for intentionally shelving applications to extend provision "for months and years," while also accusing the colonial government of deliberately keeping the "backstreets," predominantly African-populated areas, as "dark as Erebus."[30] These articles were nearly unanimous in demanding additional public investment to reduce charges and extend power sector capacity. Dove, an odd one out, instead recommended privatization.[31]

DOVE'S PRIVATIZATION PROPOSAL

It is ironic that the editors of the *Nigerian Chronicle* titled Dove's article "What People Think of Our Electric Light" because the piece was just one man's opinion, informed by his take on events abroad during this period.[32] This title is not surprising because members of Lagos's elite had a larger-than-life impression about their role in the colony's development. Emmanuel Ayandele notes that Lagosian elites used their wealth, literacy, and professional qualifications to position themselves as "the hope, the asset and the catalyst of a new Nigerian society" and often spoke or wrote as the voice of the people.[33] This group indeed laid the foundation for the "awakening of racial and political consciousness" in Nigeria, alongside later mobilizing popular protests and political movements before the Second World War.[34]

Dove had significant concerns regarding public-sector electrification. First, he accused the colonial government of supplying electricity primarily to improve the "luxur[ious lifestyle] of its officials." Since the government was also subsidizing energy consumption in official residences, as we shall see, the cost was transferred to other consumers, and, consequently, charges were exorbitant. The price of electric current in 1909 was 10 pence per kilowatt-hour (kWh), while the average daily wage of an unskilled African worker in Lagos was 13 pence.[35] To put this into context, two 40-watt incandescent bulbs left on for about twelve hours—that is, from early evening until the following morning (the hours of darkness in the tropics)—would consume approximately 1 kWh daily, costing more than two-thirds of the daily income of an unskilled worker.[36] Such an exclusionary high price meant only wealthy Europeans and middle- and upper-class Africans could realistically afford electricity.

Dove further argued that the colonial state deliberately stifled the nascent electrical energy industry by refusing to meet demand. One might assume that the state's reluctance to extend electricity supply was linked to the broader history of economic austerity in early twentieth-century colonial Africa. However, this was not the case. As mentioned, the Lagos treasury was healthy thanks to its customs receipts.[37] Dove hinted that the state's reluctance was because of high annual expenses. He quoted £10,648 as the average yearly cost of Lagos's electricity generation and supply. Yet, elsewhere, the total annual revenue from electricity supply as of August 1909 was only £5,760.[38] Dove contended that high annual electricity bills were a result of wasteful government spending. As such "the cost of lighting the public streets [the only public good aspect of electricity] so bitterly complained about" was no more than 25 percent of total government spending on electricity. For emphasis, he claimed the "heavy expenditure" (the other 75 percent) could be explained by "a glance at the Official quarters," which were excessively lit. Moreover, government officials, as part of their benefits, did not pay electricity bills and were exempt from the indirect taxes that funded electrification.

Dove instead recommended private-sector management. He believed the Lagos undertaking would become profitable if government officials paid for their share of energy consumption. However, he did not trust the government to implement policies that would jeopardize the lifestyle of colonial officials. With private-sector management, he assumed government spending would decrease, electricity would become affordable and, importantly, "poor overtaxed citizens" would be financially reprieved. Finally, surplus revenue would become a "financial asset" allowing companies to extend electricity supply to mainland places such as Ebute Metta, and in doing so, electricity could connect the "mass of the people."

PRIVATIZATION'S PITFALLS

Dove's privatization proposal faced at least two pitfalls. The first was his assumption that privatization could immediately meet demand because the power station was "thoroughly equipped to supply the whole of the city of Lagos, public streets and private houses included ... without any addition to the plant." His assumption was half true. A report published in August 1909 confirmed that the Lagos undertaking was operating at fifty-percent capacity. However, this was because of limited distribution switchboards.[39] So, while Dove was correct that the undertaking's capacity was underutilized, it was not possible that the plant could rapidly supply every street and home. For instance, assuming everyone could afford electricity, a total capacity of 120 kW would have been insufficient to serve a growing population estimated at 53,299 in 1909.[40] Government or a company would have to extend capacity and build more distribution stations to supply the entire city. This in fact happened in 1912 when the Public Works Department increased capacity to 300 kW, and in 1923 when a central power station of 3,600 kW capacity was opened.

A second, more critical pitfall was Dove's assumptions about the implications of privatization on the economic life of the average Lagosian. Large-scale electrification was expensive, and its colonial-era "shareholders" were not motivated by public interest. Aside from maritime commerce, there were few obvious incentives for private investors in Lagos in 1909. Lagos was a non-mining and non-industrial colony, relatively scarce in resources and plants—until the 1940s, when firms like the United African Company (owned by the British firm Unilever) set up factories there.[41] Since the time horizon needed to construct electricity infrastructure is decadal, it would have been difficult for private firms to make returns in the short run without raising electricity rates beyond the reach of most residents.[42]

Herein lies a dilemma: was Dove self-centered, or did he genuinely believe privatization was in the public interest? Dove's biography suggests he could afford the price increases privatization might entail. Genealogy research indicates he was a descendant of the liberated Africans.[43] He was born in Freetown, Sierra Leone, in 1863, lived for 85 years, and died in London in 1948.[44] During his lifetime, he lived in Freetown, where he was a member of the Freetown City Council and deputy mayor, moved to Lagos, and spent time in the Gold Coast and London.[45] A successful lawyer and prominent businessman, he was one of the shareholders of the *African Times and Oriental Review,* the pan-Asian-African journal.[46] Dove also floated shares in the Anglo-African and India-Rubber Trading Company Ltd in association with J. S. Sawrey, a London-based company promoter, among other enterprises.[47]

Dove, his family, and other members of the early twentieth-century West African nationalist elite lived a life of comfort under colonial rule. Most were Anglophiles and were regarded as Black Englishmen who took pride in their mastery of the English language, culture, and sartorial choices.[48] When they led protests and political movements, their aim was not to dislodge the foundation of colonial rule but to achieve self-determination within that system. Historians of West Africa, including Ayandele and Adu Boahen, have derided early twentieth-century African elite as "deluded hybrids" and self-serving "nationalist petit-bourgeoisie."[49] However, other Africanists have positively appraised their legacies, with Olúfémi Táíwò going as far as regarding them as "African apostles of modernity."[50]

Despite Dove's affluence, it might be unfair to accuse him of being *intentionally* self-serving. His 700-word article was peppered with references to the "public," "mass," and "poor." His belief that a private company could ensure "proper supply at moderate cost to the public and free from any further taxation" was not unreasonable if one considers his self-declared "careful study" of how electricity was managed in Europe. Although Dove provided no specific detail on his research, later scholars—notably Thomas Hughes—demonstrated that in Britain and Germany at least, at the turn of the twentieth century, private energy firms successfully coexisted alongside publicly funded utilities.[51] That said, such firms were highly regulated to ensure electricity supply was stable, accessible, and affordable.

Dove's comparison of the electrical question in colonial Lagos with that of Europe is akin to comparing apples and oranges. In African colonies where the private sector funded and managed power plants, electricity was not considered a public good. This situation continued throughout most of the colonial period until public electricity commissions and corporations were created to bring some measure of development and welfare to colonial subjects. Even then, investment was often guided by extractivist goals. In South Africa, two private monopolies, Victoria Falls and Transvaal Power Company Limited and the Allgemeine Elektrizitäts Gesellschaft (both financed by German banks), generated electricity principally for diamond mining.[52] Electricity remained a private good in South Africa until the government established a national Electricity Supply Commission (Eskom) in 1923.

In any case, Dove's privatization plan was dead on arrival for three reasons. First, a motive for electrification was the improvement of colonial lifestyles. Even if privately run, it is logical to assume government would have continued to subsidize the consumption of its officials. Consequently, public spending on electricity would have remained high. Second, Dove's thesis remained an exception to the norm throughout the colonial and early post-colonial periods. Leading African nationalists in Lagos and elsewhere consistently pushed for additional public investment and, in many cases, the subsidization of utilities well into the 1980s.[53] Lastly, with the benefit of hindsight, we know that the colonial government was seemingly unwilling to privatize the Lagos undertaking. Private sector interest was kindled around 1929 when the Lagos plant became profitable. However, government refused to sell off the power station.[54] A detailed analysis of this renewed privatization attempt is beyond the scope of this chapter. However, it suffices to state that the decision to retain the Lagos undertaking under government control may have been influenced by events in Britain, where nationalization of electricity infrastructure had begun in 1926.[55]

DOVE'S IDEAS IN TWENTY-FIRST-CENTURY NIGERIA

The nationalization of the British electricity industry during the twentieth century was not a unique experience. From the 1880s to the 1970s, global electrification was characterized by varying degrees of public control. At one end of the spectrum was the Soviet Union, where electrification was encouraged by socialist ideologues.[56] Some European states allowed private-sector participation but regulated the industry, as in Germany.[57] In the United States, usually the archetype of free market economies, the influence of the state on energy investment in this period is unmistakable.[58] For instance, the Federal Water Power Act of 1920 codified the federal government's power over hydroelectric resources. The act established the Federal Power Commission—which, among other things, regulated electricity tariffs.[59]

However, in response, in the United States electricity industry undertook a long and expensive public relations campaign to persuade the public of the benefits of private electricity provision.[60] More widely influential beyond the United States, while there had been preceding acts of liberalization, the overt public ownership of many state-owned industries were demonstratively overturned by powerful free-market advocates and institutions from the late 1970s onward, including Margaret Thatcher, Ronald Reagan, the World Bank, and the International Monetary Fund (IMF)—all in a sense victorious in the wake of Communism. Since then, developing countries, including Nigeria, have endured several attempts to roll back state control. Privatization debates were encouraged by the World Bank/IMF when Nigeria adopted Structural Adjustment programs in 1986. Around one hundred public enterprises were privatized, but the electricity industry survived due to its strategic significance to the state.[61] In 2005, the Electric Power Sector Reform Act was passed, which created an independent agency to regulate the industry and spur private-sector participation. The electricity industry was partially privatized in 2013.[62] Private firms are allowed to generate and distribute, while the state continues to control transmission.

In recent times, electricity reform advocates have argued that complete privatization is required to improve the efficiency in the Nigerian electricity supply industry.[63] The challenges they identified are comparable to those described by Dove. Like him, economists and other energy policy ideologues argue that the Nigerian electricity supply industry (especially the transmission sector) is not driven by profitability because it is dependent on state funding. Nigerian power stations currently operate at less than fifty percent of their capacity, as in Dove's time, though this is due to the power losses caused by a dilapidated transmission infrastructure. Nigerian plants have an installed capacity of about 12,500 megawatts. However, an average of only 6,000 megawatts is regularly supplied to distribution companies, and there are many days when the national grid is overloaded and no energy is supplied.[64]

In addition, there are overlaps between Dove's justifications for privatization and present debates. These include a desire for a pricing regime that reflects market prices, shareholder accountability, and reduced government spending. As in Dove's time, today low energy tariffs are considered to be in the public's interest, and it is argued that privatization will be to their detriment. Although electricity has been produced for over 120 years in Nigeria, the industry is not yet mature. The country needs great investment to meet its estimated 30,000-megawatt energy requirement.[65] Yet a recurring complaint promulgated by private energy companies is that state-regulated charges are too low for them to break even and attract additional investment.[66] Meanwhile, many Nigerians already feel financially overburdened and attempts to increase energy rates have triggered protests.[67]

Lastly, just as in 1909, a cross-section of the Nigerian population still believes that the state should manage the electricity industry. There is a growing desire for political interventions via investments, loan guarantees, and additional regulations, among other things, while a few analysts have gone as far as recommending renationalization.[68] As such, the debate that Dove raised remains pertinent in twenty-first-century Nigeria.

CONCLUSION

Dove's 1909 article is a novel publication by an African that advances the argument for the privatization of a public electric utility. Dove appeared to have genuinely believed the private sector was better positioned to protect public interests and to guarantee equitable access to electricity. On the contrary, this chapter argues that his privatization proposal would have had the unintended consequence of barring the average African from accessing electricity. His proposition was counterproductive for colonial Africa, considering the extractive nature of the private sector during this period. Nevertheless, this chapter argues that Dove's ideas were decades ahead of their time in showing some significant overlaps between his 1909 analysis and ongoing debates today about the political economy of Nigeria's electricity industry.

Within the context of this volume, this chapter provides a lesser-known African perspective on the management of the colonial energy industry during the early twentieth century. The historiography of electricity in Nigeria in particular is growing.[69] However, much of the literature has analyzed colonial and post-colonial economic developments. African perspectives on electrification, their perception of technologies, knowledge of energy management, and sundry issues have been rarely explored.[70] Dove presents an African's conception of energy privatization in an era when African perspectives have previously been assumed to be outside of wider debates about the political economy of energy.

FREDERICK WILLIAM DOVE
"What People Think of Our Electric Light" (1909)*

The Sister Colony of Sierra Leone is proposing to have the Streets of Freetown city lited by Electric power. Opinions seem to be divided and a correspondent of the Weekly News advised that the City Councillors ought to

---

*F. W. Dove, "What People Think of Our Electric Light," *Nigerian Chronicle*, 12 March 1909, 6–7. The orthographic variants appear in the original.

"What People Think of Our Electric Light" (1909)

take a warning from Lagos. In a subsequent issue of the same paper Mr. F. W. Dove wrote.

> That the installation of electric energy in Lagos is Governmental and therefore official and as that Government is responsible to no Shareholders but have practically an unlimited exchequer to drawn from, it does not worry itself in considering public interests and how far the undertaking could be developed for the public good but contents itself chifly in supplying the minimum of energy from the handsome and well equipped plant at its disposal at tee maximum of cost for Official purposes and such of the Mercantile firms and a few Native gentlemen who could afford to pay with no consideration for the mass of the people and public interests, in brief its charges for the private installation and supply indicate that Electric Lighting is intended to be a luxury in Lagos and as such only the Officials and well to do should enjoy it, and from my observation on the spot, I can come to no other conclusion, especially as the plant though not of the most modern type of machinery adapted for Electric Lighting purposes, is thoroughly equipped to supply the whole of the City of Lagos, public streets and private houses included, with a much more powerful than that at present supplied, an assertion confirmed by the proposal to connect Ebute Metta with the generation station, for the purposes of lighting Official quarters etc., at the former without any addition to the plant.

> It must be obvious to the man in the street with the slightest idea of the cost of Electricity that the cost of lighting the public streets of Lagos so bitterly complained of from time to time, could not be responsible for the expenditure of more than a fourth part of the £ 10,648 said to be annually expended for generating purposes in that City and for an expenditure of such fourth part a much better light covering a greater area could be supplied, but a glance at the Official quarters etc., is sufficient to explain away this heavy expenditure; unfortunately, however, I have not before me the annual revenue derived from the distribution of Electricity in Lagos and what portion of it is contributed by the Officials to enable me in arriving at the approximate figure the inhabitants of that Colony are called upon to contribute through indirect taxation for the luxury of its officials, a luxury which is a financial asset to that Colony if worked on proper lines and in the hands of a Private

Company. From the above it must be evident to your readers that the present unsatisfactory supply of Electrical energy in Lagos is due to the fact that it is under Goveenment control and it would be a suicidal policy, if at any time the City Council by refusing to vary or modify its agreement with Mr. Le Messurer to permit of arrangements already concluded by him for a proper supply at a moderate cost to the public and free from any further taxation being carried out should enter into negotiations with the Government to erect a plant here which could have but one result and that is further taxation of the mass for the benefit of the Officials as has been clearly demonstrated in re the "Water Supply of Freetown" the cost of construction and upkeep from which the Colonial Government has carefully exempted itself to the detriment of the poor overtaxed Citizens.

My observations are based from a careful study of the Electrical questions at Lagos and in Europe during the greater part of the last twelve months and are submitted with a view to dispelling any wrong impression that the extract from the "Lagos Standard" may have created as to the superiority of Electric over all other lights for public use and that such superiority could be maintained with no extra burden when controlled by other than the Government.

9 Gender, Food, and Vernacular Energy in Moussa Travélé's "Three Rapid People"

LAURA ANN TWAGIRA

IN THE FIRST DECADES of the twentieth century, food and its relation to the energy of laboring people were central concerns for West African farming societies. This preoccupation was recorded in regional folk stories such as "Three Rapid People," a tale collected and translated for publication in 1923 by Soudanese interpreter and scholar Moussa Travélé.[1] In this story, situated in the broad Mande cultural region, which stretches across a wide territory from Senegal and The Gambia to Guinea, Mali, Burkina Faso, the Ivory Coast, and Sierra Leone, the narrator tells of impossibly rapid food production and preparation via the labor of three unnamed characters: two brothers and the wife of the older brother. It is a story of human energy. The tale is set during the farming season, and because of the seasonal rains, the younger brother comically slips in the mud on his way to the fields. Thankfully, he is able to quickly save a basket of seeds from overturning. Soon after, the young farmer's older brother spies an antelope and shoots it. His accomplishment is to skin and prepare the animal before the musket ball can even reach the animal, thereby stopping the shot from spoiling their meat. When the two men return home, the wife of the older brother quickly prepared a meal, according to Travélé, a feat that would normally require four hours of work.[2] Astonishingly, the meal is ready in a matter of minutes. While the tale includes fantastic elements, the physical labor of each character is specifically highlighted, drawing together vernacular understandings of food production and preparation, the exertions of human labor, and the division of labor. More broadly, it suggests that the history of energy is profoundly gendered.

For both men and women, this energy story showcases the expenditure of human force. While it may seem to suggest a desire for the alleviation of physical work, the tale is in fact a celebration of complementary human effort. In the wider Mande region, the physically taxing aspect of men's agricultural labor, in particular, was honored with the nomination of a champion farmer during annual farming rites.[3] The younger brother, the key farming figure in the story, is charged with producing an essential grain, millet, while his sister-in-law transforms it into a meal that will fuel the two men's labors.[4] When it was recorded in the early twentieth century, listeners to the tale would have also understood that the wife's efforts were meant to feed a much larger household. The vernacular energy thinking showcased in this popular Soudanese story elevates millet, in its cooked form, as the nutritious fuel of the community. Millet was essential to human life and was produced through the physical exertion of both men and women.[5]

In the early 1920s, when "Three Rapid People" circulated in popular Mande culture and was also transcribed for a French reading public, colonial scientists were increasingly studying the caloric intake of the colonial workforce in an effort to standardize food rations across West Africa.[6] While these scientists shared the Soudanese emphasis on work in relation to food and human energy, the French emphasized quantification and standardization within an imperial political framework. Daily diets were measured and these numbers extrapolated into statistical evaluations of a minimum level of functional subsistence. Food was subsequently rationed for laborers on public works, soldiers, and other colonial workers. These workers and their labors were not glorified by the French. Indeed, in later decades French officials increasingly depicted local populations as chronically undernourished and impoverished as a result of their regional diets, all the while ignoring the devastating and long-lasting effects of conquest on food stores. Moreover, French scientists primarily studied the health and diet of laborers on French sites where it was the colonial ration that failed to fuel the workers. Yet, the French colonial narrative about the diets of France's colonial subjects was one critical of their meagre energy production (both in terms of worker output and food production). It was not a narrative that attributed much agency to the Soudanese. By contrast, the only measurement noted in the folktale—that of time—was a means to celebrate human achievement and the extraordinary production of human energy and nutrition.

While the ability of farmers to produce food was a pressing political issue for the colonial government, it was an existential matter for regional residents. Incessant colonial demands for the export of agricultural goods during World War I served only to exacerbate concerns over food supply in a region that was prone to intermittent drought. Moreover, in these same years and the following decades, the colonial government harvested wood on a massive scale to fuel steamship and rail transport, as well as to power imported industrial machinery, leaving less fuel for women's cooking fires.[7] In this context, local food production was something to be celebrated. While French scientists measured calories and rations in an effort to bolster the extractive economy and directed the consumption of wood fuel to industry, Soudanese vernacular energy thinking emphasized human potential.

Unstated in this folk tale but central to its energy story is *nyama*, a natural element but also a supernatural force that extraordinary people may harness. The linguist Charles Bird has translated the concept as "the energy of action" and further explains that it is "the necessary power source behind every movement, every task."[8] For example, *nyama* propels the musket ball in the tale, but it is also the source of the hunter's ability to beat the ball to his prey. In regional oral traditions the hunter figure is a recurrent supernatural hero. He is knowledgeable about the wider natural and political world because of his travels in search of hunting grounds. He is also called upon for his advice and to serve as a protector. As a figure of vernacular energy thinking, the hunter learns to harness *nyama* in order to capture prey and prepare medicines—some supernatural—from products found in the bush. As art historian Patrick McNaughton explains, collective *nyamaw* "are the energies that animate our world."[9] It is a form of energy that is potentially helpful to humans but also possibly deadly and as such necessitates human effort to manage or control it.[10] Besides hunters, only exceptional people are able to harness *nyama*, and they include healers, sorcerers, and powerful rulers in epic tales, as well as blacksmiths who manipulate fire and earthly mineral elements in their work. Women specialists such as potters harness ritual energy in their work with earth, fire, and water. We can only speculate on the tale's early twentieth-century interpretations, but one reading might be that women similarly control *nyama* to cook and ought to feature alongside hunters in popular heroic tales. Specifically, women employ fire to transform earthly elements, even in times of scarcity, into cooked meals.[11] In short, women's *nyama* ensured food and fuel in uncertain times.

Nyama is additionally expressed and channeled in the spoken word, such as in the telling and retelling of "Three Rapid People." Specialists of the ritual and entertaining spoken word inspire emotion and incite action in the intended listener.[12] In the case of this story, *nyama* boosts the energy potential of the three rapid people *and* in the listeners of the tale. As political scientist Cara Daggett argues, energy is a "slippery word"; it is something more than a substance to be measured scientifically in terms of its productive and hence economic value; it is deeply entangled with understandings of work, social change, and communal values.[13] In Soudanese culture, human labor is more than physical exertion; its transformative action is noble, something to be mastered, and when it comes to producing food, something requiring supernatural ability.

Moussa Travélé was an interpreter who worked for the colonial government in French West Africa when he collected and translated the tale. It was published in *Proverbes et contes bambara,* a book containing an opening ethnographic analysis of Bambara and Malinke societies, both part of Mande culture, along with a rich corpus of other regional tales and proverbs.[14] In 1923, colonial rule was transforming the region in which Travélé collected his material, and he was particularly concerned with documenting pre-colonial life before it was potentially lost.[15] At the same time, he recorded popular critical commentary on colonial rule, such as the proverb: "The rabbit looks like the donkey (because of their ears), but it is not his son. (That is to say, for example, the Blacks who want to be like the Whites.)" (*Nsonsan bolé fali fè, nk'à dén tè.—Le lièvre ressemble à l'âne (à cause de ses oreilles), mais ce n'est pas son fils. [Se dit, par exemple, des Noirs qui veulent être assimilés aux Blancs.]*)[16] His selection of material for *Proverbes* similarly expressed a number of concerns for life under colonial rule and for the future of Soudanese society. For example, his 1923 ethnography and selection of tales reinforced the idea of male authority in domestic life during a period of social upheaval. His collection of stories also showcased popular interest in the pleasures of eating and the social importance of agricultural labor, but also anxieties about food shortages, which were exacerbated under colonial rule. So, while he presented his corpus of oral materials as representative of pre-colonial life and set in the past, it suggests a more dynamic vernacular conversation about adapting to ongoing changes.

Travélé's collection of early twentieth-century oral literature is incredibly valuable as one compiled by an African author. It also requires careful reading. As literary scholar Laurence Porter explains, oral stories collected by educated Africans in the colonial era lost some meaning in the process of translation.[17] Indeed, Travélé's French translations vary from the original Bambara performances and his own Bambara transcriptions of the tales. For example, Travélé acknowledges the element of repetition in the performances of some of the stories he collected, but he quickly passes over such moments by simply indicating to the reader that the same formula was spoken by another character, or that the same pattern of action was repeated. If the spoken word conveyed *nyama*, it follows that repeated words and suggestions of repeated actions augmented the energy potential of a performance.

What is remarkable is that Travélé offers a Bambara version for each proverb, riddle, and tale alongside his own French translation. This inclusion of a Bambara version of the oral material allows those conversant in the language to study a text that is closer to the spoken versions circulating during the period, even if its performative aspects have been omitted.[18] Indeed, Travélé's aim was that French readers would study the Bambara texts as a means to understand Soudanese cultures and languages. He had previously published a Bambara grammar guide, *Petit manuel français-bambara* (1910), and a Bambara-French dictionary, *Petit dictionnaire français-bambara et bambara-français* (1913). Both books included a few folk stories, and the *Petit manuel* also included proverbs, suggesting he had collected literary material over several years.

The tales that Travélé published in *Proverbes and contes bambaras* were most likely composed prior to the 1920s, and he may have heard many of them growing up in the region around the turn of the century. In interpreting this story as a source for vernacular energy thinking, I draw on scholars who have established that rumors, folk tales, myths, and other oral cultural productions are valuable sources for historical study. For example, historical anthropologist Roderick McIntosh has argued that Mande social and environmental memory is embedded in oral traditions passed down over generations. Moreover, key figures in these stories—such as hunters and blacksmiths—interpret environmental changes and impart lessons for understanding the landscape and coping with ecological crises.[19] The recurring trope of the hunter figure is significant here for its larger cultural resonance.[20] Food production, be it farming, hunting for meat, or cooking, is another formulaic element in this story, and as Luise White argues, such scenarios provide recognizable "images into which local meanings and details are inserted by their tellers."[21] In this way, the folk tale "Three Rapid People" offers a window into a regional conversation about food as fuel, the value of physical labor, and the gendering of energy work.

In Travélé's original Bambara transcription of "Three Rapid People," the two brothers are heading to work in millet fields. This is the same grain that the older brother's wife prepares for the evening meal. Although Travélé might not have thought his French audience needed this information, the grain is central to rereading the tale as an energy story. In fact, a popular Bambara expression connects men and women to different moments in the preparation of millet: "It is for the husband to supply the *toh* [a dish prepared with millet] and for the woman to supply the sauce."[22] Although women prepared the whole meal (the *toh* and the sauce), men symbolically provided the essential grain from their field labors. In fact, men's millet cultivation was also celebrated with the performance of physically taxing masked dances and the designation of champion farmers.[23] Even though women worked to weed the millet fields and to transport the harvest, the idealized gendered division of labor in the late nineteenth and early twentieth centuries linked men with the socially significant task of millet production. *Toh* was the filling portion of the meal that fueled long hours of male labor in the fields. Indeed, the phrase *toh fanga* draws the dish together with the idea that it gives force or power, *fanga*.[24] For a Bambara or Malinke audience, the millet in the story of the "Three Rapid People" was a familiar and important cultural reference. Millet fueled human labor and society, both materially and symbolically. Although the tale reinforced the specifically gendered tasks associated with the grain, it also showcases women's role in producing *toh*. The human labor of producing energy (or energy work) was both men's and women's work.

Moreover, "Three Rapid People" presents an energy story focused on the household; it is certainly an idealized depiction featuring full grain stores filled through family cooperation and is seemingly absent of regional politics. It contrasts another folk tale about food production collected by the colonial administrator Charles Monteil in the last years of the nineteenth century. In his translation of "The Hyena," the protagonist hyena finds a magical cooking pot that produces meals on command. After sharing the object with an elderly female host, she gifts it to the regional ruler, rather than using the pot to feed her guest. This depiction of food production is not at all idyllic and is even potentially criticizing the elder woman for her mismanagement of food resources or the extent of her control over them. Still, at the end of the story the hyena retrieves the pot through cunning.[25] One way to read this tale is that it alluded to a contested political tribute sent to ruling elites or even the raiding of food stores by regional rivals. Historian John Cropper, who studies energy politics in the Senegal Valley, cites a similar Wolof story about a starving man who finds a bowl that magically produces millet. Unfortunately for him, the king confiscates the bowl. Cropper offers the interpretation that grain harvests are unreliable and require a resilient agricultural system.[26] An additional reading of the Wolof tale is that it was a popular critique of elites who abuse their power. As Cropper argues, control over grain supplies for Wolof rulers over several centuries was a means to control energy production and its infrastructure. It was accomplished largely through capturing the labor of "households, compounds, and agricultural workspaces, [what he calls] organic refineries where Wolof populations converted raw materials and anthropogenic landscapes into metabolic, mechanical, and thermal energy."[27] In the Senegal Valley in the late seventeenth and eighteenth century, controlling this people-powered energy infrastructure was increasingly coercive, meriting such popular rebuke as in the tale.

In the story collected about a century later east of the Senegal Valley by Monteil, the hyena tricks the ruler into trading the pot for a magical knife that strikes him. The surprise attack allows the hyena to recover the pot and flee. At the time the tale was collected by Monteil, it was perhaps locally interpreted as a critique of French conquest and the subsequent exactions on local agricultural labor. Taken together, "The Hyena" and "Three Rapid People" imagined an agricultural world where energy production was plentiful and controlled locally rather than by an intrusive state.

"Three Rapid People," in particular, centered the power and agency of individual people as producers of energy. Early twentieth-century listeners to this tale also might have identified the hunter with speed. By the 1870s regional hunters and soldiers had access to firearms, which would have made the act of killing an antelope faster than in previous decades. Indeed, the insurgent armies of Samori Touré employed locally manufactured firearms.[28] Listeners in the 1920s also would have been accustomed to the practice of hunting by firearm, as is suggested by the character's use of a musket, yet the speed of the gun's shot was still remarkable. At the same time, the hunter in the story does not wish for the ball to ruin the meat. Since arms were also produced in local forges and workshops, the hunter may have been doubly associated with the *nyama* of both the hunter and the blacksmith. Unstated is the history of gun use by the French forces of conquest, as well as the soldiers under Samori Touré's command. They were notorious for violence and wielded firearms to expand Touré's land claims and labor force at the end of the nineteenth century.[29] Such immoral productions and deployments of *nyama* were widely remembered and critiqued during the first decades of colonial rule. Here, rather, it is the hunter's use of a gun and *nyama* in order to produce meat for his brother and wife that is highlighted.

Soudanese women who listened to this tale in the 1920s no doubt fully understood that they, like the two men in the tale, were also producers of energy. Indeed, like the male farmer, women labored in the millet fields and carried out specific women's work in the fields. They were also aware of the high social value placed on productive physical labor. However, women likely more closely identified with the physical work of transforming millet into a meal. As historian Eugenia Herbert has argued, societies across Africa conceptualized technologies of transformation as central to not only social reproduction and daily subsistence, but also ritual life.[30] Herbert's argument focuses on the gendered language and power of metal working and pottery production, but her argument is also instructive regarding the production of food. Women transformed millet into meals that gave power and, like metalworkers and pottery producers, engaged with dangerous elemental forces such as fire to do so. Women exerted energy growing and collecting ingredients for the sauce and collecting firewood, but the emphasis in "Three Rapid People" on the final meal focuses on women's labor in the preparation of *toh*—despite women's wide cultural association with "the sauce." Indeed, women's labor was a central concern for rural communities in the 1920s as increasing numbers of women protested against their labor conditions in marriage, as well as other domestic concerns, by simply leaving their husbands. Confused colonial authorities faced complaints from men and sought to manage a growing runaway-bride crisis.[31] The loss of women's labor shed light on its value and perhaps influenced the telling of a tale that honored women as exceptional cooks.

At the end of the tale, the narrator asks the audience, "Who is the fastest of these three people?" In posing the question, the narrator also seems to be asking who is the most astonishing. The final question might also be whose human labor or force is the most essential to the shifting agricultural communities of the early twentieth century. Women in the audience may have likely chosen the woman in the tale. Rural Soudanese women weeded together in the fields, pounded millet together before it was cooked, and socialized while doing other women's work. While they may not have always labored in harmony, they did share the common obligation and the prestige of women's work, much of which required both speed and force. Women's specifically gendered labor (cooking), like the men's gendered labor, centered the role of the human body in producing the necessary energy for human life. Indeed, millet in the story appears to give men and women the ability to perform their labor fantastically fast. However, this fuel was only fully produced through the complementary physical labor of men *and* women.

A mid-twentieth-century audience for this tale would likely have interpreted the energy story differently. For example, male farmers were still celebrated for their labors in the fields, but increasingly they embraced the use of the plow powered by animal energy (and even the tractor). When plows were first introduced in the 1920s, some men refused to eat a meal made from the millet grown with the aid of a plow. The marks it made in the earth seemed only to scratch the dirt. How could it produce millet that would make for filling food?[32] It also undermined the symbolic import of men's physical labor.[33] Here, the male energy story shifted when regional blacksmiths began to modify imported plows. They gained local meaning and masculine associations.[34] Women could also listen to this story with an awareness that their labors were still physically taxing but newly altered by a shift in their energy regime. Around mid-century, women began using metal cooking pots. These new pots used less wood fuel to produce a meal, and they even cooked faster than older clay ones.[35] In both cases, the human energy component was altered, but the production of millet and *toh* as fuel remained a key concern. But women's labor still transformed the grain from the fields that today carries the name "for life" (*ka balo*).³⁶ Travélé's vernacular energy story was not something received within a kind of timeless energy-using society. It was shared and discussed within communities who, under colonial rule, engaged with the shifting technological possibilities of plows, tractors, and metal pots and their meanings for both human labor and food as fuel.

MOUSSA TRAVÉLÉ
"Three Rapid People" (1923)*

A man and his younger brother go to the fields with the little brother carrying a basket of seeds on his head. It so happened that a strong rain had fallen the night before, and the little brother slipped. But before he fell, had had time to place the basket of seeds on the ground and to change from his bright white boubou [large top] and pants into his older boubou and pants, before finally falling.

They continue to the field and sit under a tree. A doe (called *nkoloni* [antelope]) runs by the two. The older brother takes his musket and fires at the antelope. Impatient, he grabs the doe, kills and skins it, even hanging the skin next to where he was sitting, all in one motion. He then meets the ball, takes it, and says: "You are not going to ruin my meat."

Night begins to fall, and they return home.

The wife of the older brother tells her young brother-in-law to go to the granary (for millet) and bring her some *keninke kise* [big millet] to prepare dinner.

The young man goes into the granary, fills a small basket, and gives it to his sister-in-law. He goes back and refills the basket, and again gives her the

---

*Moussa Travélé, "Trois Personne Rapides," in *Proverbes et contes bambaras (Accompagnés d'une traduction française et précédés d'un abrégé de droit coutumier)* (Paris: Librarie Orientaliste Paul Geuthner, 1923), 57–58. Translated by Laura Ann Twagira.

"Three Rapid People" (1923)

grain. The third time, she says: "Don't drop those millet grains on my *toh* (which is already cooked)."

So we must ask, which of these three people is the most rapid?

*

[Travélé's notation of "Trois Personnes Rapides," alternating between Bambara and French]

Tiè n'a dogokè bé tâ forola. Gno-si sègui bè douakè koun. O doun y'a soro san-ba nalé soufé. Douakè tiénéra san'a ka bi a li gno-sègui djigui, k'a-ka doroki diè n'a-ka kouronsi diè bo; ka doroki kolon ni kourounsi kolon don, ka soro ka bi.

Un homme et son petit frère allaient aux champs. Le panier de semence était sur la tête du petit frère. Il se trouva qu'une forte pluie était tombée pendant la nuit précédente. Le jeune frère glissa, (mais) avant de tomber, il eut le temps de déposer le panier, de se déshabiller de son boubou blanc et de son pantalon blanc ; puis il mit son vieux boubou et son vieux pantalon et tomba ensuite.

Ou târa forola k'i sigui iri koro. Nkoloni boribâto nara témé. Korokè y'a-ka marifa ta k'o tji ola ; ka koroto ka bori ka tâ sogo minè, k'o kan tiguè, k'o boso, ko sogo kè o golola, k'o glo i kéréla, ka na marifa kisé bèn, k'o minè, k'a fo : « Kana tâ n'ka sogo ratien ».

Il[s] allèrent au champ et s'assirent sous un arbre. Une biche (du nom de *nkoloni*) vint à passer en courant. L'aîné prit son fusil, tira sur elle ; dans son impatience, il alla lui-même prendre la biche, l'égorgea, la dépouilla, mit la viande dans la peau, la suspendit à son côté. Il rencontra la balle, la prit, en lui disant : « Ne vas pas abimer ma viande ».

Oulada sélé, ou târa so.

Le soir ils rentrèrent à la maison.

Korokè moso ko k'a nimâni ka don djiguina kono k'â son kéninkè kisé do, ka bo kè sourafana yè.

La femme de l'aîné dit à son petit beau-frère d'entrer dans le grenier (à mil) et de lui donner du gros mil pour faire le dîner.

Douakè dona djiguina kono ka tiéré fa k'â di moso ma, k'â fa k'â di; a sabanan moso ko : « I kana gnokisè kè n'ka tôra ».

Moussa Travélé

Le jeune homme entra dans le grenier, emplit un petit panier, le donna à la femme, l'emplit de nouveau, le lui donna; à la troisième fois, la femme dit: « Ne mets pas de grains de mil dans ma pâte (qui est cuite) ».

Ko ni mâ saba dioumèn ka tari to-ou yè ?

On demande quelle est la plus rapide de ces trois personnes ?

10 Uncertain Energy Epistemologies

William James and the Case of Mental and Moral Energy

REBECCA WRIGHT

IN AN "INTRODUCTORY" TO a 1914 republication of William James's 1906 speech "The Energies of Men," the New York–based publisher Moffat Yard & Company felt it necessary to caution that despite appearing that "the sane and simple message of this essay could not be misconstrued," a clarification was due. The essay "does not counsel all persons to drive themselves at all times beyond the limits of ordinary endurance, that it is not a gospel of overstrain nor an advocate of the use of alcohol and opium as stimulants in emergencies."[1] Republished after James's death in 1910, the fact that the essay required an additional note of clarification speaks towards the afterlife of James's address as it caught the public imagination. In the decades that followed, commentators increasingly interpreted James's essay as a manifesto celebrating energy, brute power, and domination. Read as part of James's broader philosophical project, however, far from being a simple celebration of force, James's essay was making an important statement about the relationship between energy and society. Moreover, James explored unresolved uncertainties about the scope and explanatory power of the laws of energy from the heart of a country that was increasingly the world leader in the basic and applied sciences of energy, not least in physics and engineering.

William James is hardly considered a great energy theorist. An American psychologist, philosopher, and public intellectual, he is best known as a towering figure within the history of American intellectual thought and modernist studies.[2] Based at Harvard University for most of his career, James was a pioneer in the fields of functional psychology and philosophy, becoming a leading advocate of pragmatism in the United States. He didn't spend his time theorising about fossil fuel extraction or the impact of natural resources upon modern society. Nor did he reflect explicitly upon the relatively new technology of electricity that expanded throughout his lifetime, even though allusions to these infrastructures would punctuate his work and provide models for his thinking.[3] At the heart of James's work, however, is what Sergio Franzese has defined as an "ethics of energy," a moral philosophy centred around the social organization of energy.[4] For James, energy was a social and political matter. Far from providing a deterministic account of energy and society favoured by his contemporaries, such as Henry Adams (see the introduction to this volume), James placed the responsibility on the individual "will" and society to organise energy. This energy ethic was developed throughout James's prolific writing career and his major works, such as *The Varieties of Religious Experience* (1902) as well as through his speaking engagements, which attracted widespread public attention during his lifetime. Moreover, this "energy ethic" came to define many of the ways that energy filtered into the American social imaginary in the first half of the twentieth century.

James was at the end of his career and at the height of national and international fame when he penned his speech "The Energies of Men." Earlier that year he had been teaching at Stanford University when the 1906 San Francisco earthquake devastated the city. This event had been a foundational moment for James, who experienced for the first time the power brought about by the rush of adrenaline in the face of adversity.[5] This experience would inspire him to write his speech "The Energies of Men," reflecting upon the extraordinary power humans are capable of when pushed to the extreme. The essay was first presented at the psychology club at Harvard University in April 1906.[6] By December that year he had developed his reflection into his speech "The Energies of Men," given as the Presidential Address to the American Philosophical Association at Columbia University.[7] Expecting an address dedicated to a scientific subject, the audience would have been surprised that James would have chosen to discuss a subject more familiar to the "the moralist and mind-curers."[8] That was the question of mental and moral energy. James set himself the task of finding an explanation to better understand "the *amount of energy available* for running one's mental and moral operations by."⁹ Despite, he pointed out, everyone having an intuitive grasp of this subject, no one had provided a scientific explanation as to why on some days one had more energy than others.¹⁰ Was it possible, he wondered, to find a plausible explanation for the phenomenon of "second wind" everyone intuitively experienced throughout their lives.

The fact that James turned his attention to the subject of mental and moral energy in this high-profile address is not surprising. American culture at the turn of the century was gripped with the subject of energy.¹¹ Helmholtzian materialism had introduced energy into the biological sciences influencing nascent fields, such as psychology. Taylorism and the "science of work" was increasingly becoming applied to the management of labour and the human body.¹² New technologies, such as electricity, were expanding into people's everyday life, becoming a topic of cultural fascination.¹³ Energy was also becoming heavily valorised within the cultural mainstream. Popular figures, such as Theodore Roosevelt, would cement their fame by popularising the mantra of "the strenuous life," a life of vigour and physical prowess.¹⁴ Counterbalanced with this rhetoric was an obsession with the second law of thermodynamics, as entropy became a popular trope to describe the decline of the American nation.¹⁵ Not only did the conservationist movement stress the importance of saving the nation's natural resources (its coal, forests, and wilderness resources), but alongside this there was a wider discourse about the preservation of "human energy" or "human resources." "Neurasthenia," for example, became the ailment of the period—a condition first coined by the neurologist George Miller Beard in 1869 to connote a weakness of will or the exhaustion of physical energy.¹⁶ Symptoms ranged from sleeplessness to headaches and complete mental breakdowns. This condition, Carolyn de la Peña has pointed out, drew upon a "closed-system view of human energy: the body had a limited amount of force or energy that travelled through the nerves and produced productive force."¹⁷ Many of the great thinkers of the day suffered from the condition of neurasthenia. James himself was a famous neurasthenic, often taken to long periods of depression throughout his life. At the time that James was writing, therefore, energy saturated the social, political, and cultural imaginary of late nineteenth-century America. James himself embodied this rhetoric in his own body as it frequently was overcome with exhaustion.

"The Energies of Men" could be read as an attempt to overcome the limitations of the neurasthenic body. Far from reducing the body to a closed-energy system, however, James's model of "second wind" highlighted the limitations of the energy laws. After all, James was clear in his essay that despite being a dominant physical model for explaining the universe, the energy laws broke down as they were applied to the question of "mental or moral energy." He noted, even though the "conception of 'energy'" in physics is aligned to work, "mental energy" although "absolutely indispensable to our lives," lacked any scientific quality; "it offers itself as the notion of a quantity, but its ebbs and floods produce extraordinary qualitative results."[18] This incongruity did not stop James from trying to apply the energy laws to something as nebulous as "mental or moral energy." In fact, it was the tension between the quantity of energy and its quality that fascinated James. For James, not all forms of energy were equivalent. He explained, "Our muscular work is a voluminous physical quantity, but our ideas and volitions are minute forces of release, and by 'work' here we mean the substitution of higher *kinds* for lower *kinds* of detent. Higher and lower here are qualitative terms, not translatable immediately into quantities."[19] James, therefore, pointed out that there was no simple correlation between a quotient of energy and its value as it was put to work in society.

In making this distinction, James was placing himself in opposition to energy determinists, figures such as the historian and social theorist Henry Adams, who read social progress against a direct measure of energy available in the universe. A friend of James, Henry Adams was similarly drawn to the laws of thermodynamics—in particular, the second law, using this as a schema to understand historical progress. Adams's clearest explanation of this was in *A Letter to American Teachers of History* (1910), published a few years after "The Energies of Men," which applied the second law to chart the decline of civilization.[20] Reading this tract, James, who was nearly at the end of his own life, was shocked by such a crude application of the energy laws to society. In a series of letters to Adams sent in 1910, James attacked the conceptual fallacy at the heart of Adams's treatise. Adams, he maintained, had confused "absolute physical units" with "the distribution of its effects." The laws of thermodynamics could provide no measure for historical development, because, James maintained, it was the brain that "issue[s] proclamations, write[s] books, describe[s] Chartres cathedral, etc." and as a result drives historical progress, which was minimal when "measured in absolute physical units."[21] As such, he reminded Adams of the importance of "human institutions" in driving progress; "their value has in strict theory nothing to do with their energy budget—being wholly a question of the form the energy flows through."[22]

Throughout "The Energies of Men" James presented examples drawn from his acquaintances and the biographies of "great men" to better understand how to unlock this higher form of energy. In doing so, he stressed the importance of "dynamogenic" agents, an experience, idea, or psychical practice that "will launch a man on a higher level of energy for days and weeks, will give him a new range of power."[23] As James explained, it was the external factors, "ideas" or "events," which were essential for delivering a higher level of energizing. These acted as powerful catalysts for releasing the stores of pent-up energy. Throughout his writing, therefore, James, as Franzese has demonstrated, placed the onus on the individual and society to organise the energy in the universe. In lieu of external pressure points, energy would remain untapped, unorganised, and subject to entropy. Instead, energy for James had to be situated in relation to a higher ideal or moral value which bestows upon it its value.[24] This higher end would be determined, in accordance with his pragmatist philosophy, through its ability to act upon the world to enhance or improve it. Rather than placing the moral onus on "energy," James stressed the importance of developing the physical and mental infrastructure to harness energy to socially beneficial ends.

Like all of James's work, "The Energies of Men" was a call to action. In his conclusion, James made a case to the philosophical profession to establish "a methodical inventory of the paths of access, or keys, differing with the diverse types of individual, to the different kinds of power."[25] This had ramifications, James asserted in a revised edition of the essay, for the "national economy, as well as of individual ethics."[26] In order to optimise society's energy, Franzese points out, James believed that energy would have to be disciplined at both an individual level and at a social level. On the individual level, alternative therapies, such as mind cure and yoga could play an important role. These therapies allowed people to discipline their energy; they allowed a shift "into gear energies of imagination, of will and of mental influence over physiological processes, that usually lie dormant."[27] James had tried out a number of these new age methods himself, such as yoga, although he had been disappointed with the results.[28] At the social level energy had to be organised through a collective praxis. At this stage, Franzese explained, "The organization of energy involves the constitution of collective habits, which are properly thought of as institutions. Institutions appear in this context as dynamic forms of equilibrium in the interplay among individuals' energies, captured in some collective tendency."[29] In developing an ethics oriented around the organisation of energy, James placed the responsibility on society, on the end to which energy was harnessed as the key to social advancement.

One of the clearest articulations of how James envisioned the practical application of this energy ethics would occur in one of his later essays, "The Moral Equivalent of War" (1910). Here James applied the conservation law to social action, calling for a "moral equivalent of war" akin "to the mechanical equivalent of heat."[30] A passionate anti-imperialist and pacificist, in this essay James proposed the creation of a mechanism for redirecting the surplus energy engendered within the warlike condition towards a more positive end: "a war against Nature." To do this, James proposed the creation of a national youth service who would be enlisted to "coal and iron mines, to freight trains, to fishing fleets in December, to dishwashing, clothes washing and window washing, to road building and tunnel-making, to foundries and stoke-holes, and to the frames of skyscrapers."[31]

James's energetic programme extended far beyond his own work to shape the broader imaginaries of energy in early twentieth-century America. Not only did his "moral equivalent of war" enter common parlance, indeed it remains a phrase that circulates today, but his emphasis upon an ethics of energy extended far beyond his own work to inform the reform tradition that dominated the Progressive Era. Reform rhetoric frequently turned to the necessity of building both the physical and moral architecture required to channel society's energy.[32] Within this discourse, reformers often described how socially marginalised groups, such as youth, immigrant populations and women were failing to apply their energy towards socially productive ends. "Misdirected energy" became a trope to describe the ills that plagued industrial society. Emphasis rested upon developing a more resilient culture and built environment that could control and rehabilitate society's energy. For example, reformers considered how far the built environment could be improved to better direct the energy within the social body. One example of this was the playground movement, which advocated the creation of play areas in congested urban centres and spoke about their important role in "channelling" and "disciplining" the wayward energies of youth.[33] Others called for the restructuring of social institutions to release women's energy away from the preservation of the family towards economically productive activities.[34] The writer and feminist Charlotte Perkins Gilman called for the eradication of "home" as a social institution, arguing that communal living (with rationalised cooking, cleaning, and childcare facilities) would facilitate a great diversion of women's energy towards more socially productive ends.[35] Energy, therefore, deeply informed the ways in which American institutions and the built environment were discussed and envisaged in the early twentieth century.

The adoption of energy within James's philosophical programme provides a valuable example of an energy epistemology which has emerged alongside the growth of energy systems. Although not directly referring to energy technologies, James's energy ethics offered a model for conceptualising energy and society in America in the early decades of the twentieth century. Of course, James's energy ethics was riddled with its own contradictions and conceptual fallacies, as he himself freely admitted. Within it, however, we can locate an early critique of energy determinism that would come to dominate thinking about energy and society in the years to come. James was adamant that energy's value did not lie in its magnitude but in its distribution—in its organisation within a society. In doing so, James placed the onus on society to utilise and harness energy to productive ends. Rather than offering a blind valuation of energy as a social good, it was only through social organization and the improvement of human institutions that energy could be directed towards higher ends. As we confront the challenge of decarbonization, it would be apposite to remember James's warning that it is not the amount of energy in a society at any one time that counts, but how energy is put to use that is of prime importance when determining its value.

WILLIAM JAMES
"The Energies of Men" (1907)*

I wish to spend this hour on one conception of functional psychology, a conception never once mentioned or heard of in laboratory circles, but used perhaps more than any other by common, practical men—I mean the conception of the *amount of energy available* for running one's mental and moral operations by. Practically every one knows in his own person the difference between the days when the tide of this energy is high in him and those when it is low, though no one knows exactly what reality the term energy covers when used here, or what its tides, tensions, and levels are in themselves. This vagueness is probably the reason why our scientific psychologists ignore the conception altogether. It undoubtedly connects itself with the energies of the nervous system, but it presents fluctuations that cannot easily be translated into neural terms. It offers itself as the notion of a quantity, but its ebbs and floods produce extraordinary qualitative results. To have its level raised is the most important thing that can happen to a man, yet in all my reading I know of no single page or paragraph of a scientific psychology book in which it receives mention—the psychologists have left it to be treated by the moralists and mind-curers and doctors exclusively.

Every one is familiar with the phenomenon of feeling more or less alive on different days. Every one knows on any given day that there are energies slumbering in him which the incitements of that day do not call forth, but which he might display if these were greater. Most of us feel as if we lived habitually with a sort of cloud weighing on us, below our highest notch of clearness in discernment, sureness in reasoning, or firmness in deciding. Compared with what we ought to be, we are only half-awake. Our fires are damped, our drafts are checked. We are making use of only a small part of our possible mental and physical resources. In some persons this sense of

---

*William James, "The Energies of Men," *Philosophical Review* 16, no.1 (1907): 1–20 (selected extracts). Reprinted by permission of the publisher Taylor & Francis Ltd.

"The Energies of Men" (1907)

being cut off from their rightful resources is extreme, and we then get the formidable neurasthenic and psychasthenic conditions, with life grown into one tissue of impossibilities, that the medical books describe.

Part of the imperfect vitality under which we labor can be explained by scientific psychology. It is the result of the inhibition exerted by one part of our ideas on other parts. Conscience makes cowards of us all. Social conventions prevent us from telling the truth after the fashion of the heroes and heroines of Bernard Shaw. Our scientific respectability keeps us from exercising the mystical portions of our nature freely. If we are doctors, our mind-cure sympathies, if we are mind curists, our medical sympathies are tied up. We all know persons who are models of excellence, but who belong to the extreme philistine type of mind. So deadly is their intellectual respectability that we can't converse about certain subjects at all, can't let our minds play over them, can't even mention them in their presence. I have numbered among my dearest friends persons thus inhibited intellectually, with whom I would gladly have been able to talk freely about certain interests of mine, certain authors, say, as Bernard Shaw, Chesterton, Edward Carpenter, H. G. Wells, but it wouldn't, it made them too uncomfortable, they wouldn't play, I had to be silent. An intellect thus tied down by literality and decorum makes on one the same sort of impression that an able-bodied man would who should habituate himself to do his work with only one of his fingers, locking up the rest of his organism and leaving it unused.

In few of us are functions not tied-up by the exercise of other functions. G. T. Fechner is an extraordinary exception that proves the rule. He could use his mystical faculties while being scientific. He could be both critically keen and devout. Few scientific men can pray, I imagine. Few can carry on any living commerce with "God." Yet many of us are well aware how much freer in many directions and abler our lives would be, were such important forms of energizing not sealed up. There are in everyone potential forms of activity that actually are shunted out from use.

The existence of reservoirs of energy that habitually are not tapped is most familiar to us in the phenomenon of "second wind." Ordinarily we stop when we meet the first effective layer, so to call it, of fatigue. We have then walked, played, or worked "enough," and desist. That amount of fatigue is an efficacious obstruction, on this side of which our usual life is cast. But if an unusual necessity forces us to press onward, a surprising thing occurs. The fatigue gets worse up to a certain critical point, when gradually or suddenly it passes away, and we are fresher than before. We have evidently tapped a level of new energy, masked until then by the fatigue-obstacle usually obeyed. There may be layer after layer of this experience. A third and a fourth "wind" may supervene. Mental activity shows the phenomenon as well as physical, and in exceptional cases we may find, beyond the very extremity of fatigue distress, amounts of ease and power that we never dreamed ourselves to own, sources of strength habitually not taxed at all, because habitually we never push through the obstruction, never pass those early critical points.

When we do pass, what makes us do so?

Either some unusual stimulus fills us with emotional excitement, or some unusual idea of necessity induces us to make an extra effort of will. *Excitements, ideas, and efforts*, in a word, are what carry us over the dam.

In those hyperesthetic conditions which chronic invalidism so often brings in its train, the dam has changed its normal place. The pain-threshold is abnormally near. The slightest functional exercise gives a distress which the patient yields to and stops. In such cases of "habit-neurosis" a new range of power often comes in consequence of the bullying-treatment, of efforts which the doctor obliges the patient, against his will, to make. First comes the very extremity of distress, then follows unexpected relief. There seems no doubt that we are each and all of us to some extent victims of habit-neurosis. We have to admit the wider potential range and the habitually narrow actual use. We live subject to inhibition by degrees of fatigue which we have come only from habit to obey. Most of us may learn to push the barrier farther off, and to live in perfect comfort on much higher levels of power.

Country people and city people, as a class, illustrate this difference. The rapid rate of life, the number of decisions in an hour, the many things to keep account of, in a busy city-man's or woman's life, seem monstrous to a country-brother. He doesn't see how we live at all. But settle him in town; and in a year or two, if not too old, he will have trained himself to keep the pace as well as any of us, getting more out of himself in any week then he ever did in ten weeks at home. The physiologists show how one can be in

"The Energies of Men" (1907)

nutritive equilibrium, neither losing nor gaining weight, on astonishingly different quantities of food. So one can be in what I might call "efficiency-equilibrium" (neither gaining nor losing power when once the equilibrium is reached), on astonishingly different quantities of work, no matter in what dimension the work may be measured. It may be physical work, intellectual work, moral work, or spiritual work.

Of course there are limits: the trees don't grow into the sky. But the plain fact remains that men the world over possess amounts of resource, which only very exceptional individuals push to their extremes of use.

The excitements that carry us over the usually effective dam are most often the classic emotional ones, love, anger, crowd-contagion, or despair. Life's vicissitudes bring them in abundance. A new position of responsibility, if it do not crush a man, will often, nay, one may say, will usually, show him to be a far stronger creature than was supposed. Even here we are witnessing (some of us admiring, some deploring—I must class myself as admiring) the dynamogenic effects of a very exalted political office upon the energies of an individual who had already manifested a healthy amount of energy before the office came.

[...]

Conversions, whether they be political, scientific, philosophic, or religious, form another way in which bound energies are let loose. They unify, and put a stop to ancient mental interferences. The result is freedom, and often a great enlargement of power. A belief that thus settles upon an individual always acts as a challenge to his will. But, for the particular challenge to operate, he must be the right challengee. In religious conversions we have so fine an adjustment that the idea may be in the mind of the challengee for years before it exerts effects; and why it should do so then is often so far from obvious that the event is taken for a miracle of grace, and not a natural occurrence. Whatever it is, it may be a highwater mark of energy, in which "noes," once impossible, are easy, and in which a new range of "yeses" gain the right of way.

We are just now witnessing—but our scientific education has unfitted most of us for comprehending the phenomenon—a copious unlocking of energies by ideas, in the persons of those converts to "New Thought," "Christian Science," "Metaphysical Healing," or other forms of spiritual philosophy, who are so numerous among us to-day. The ideas here are healthy-minded and optimistic; and it is quite obvious that a wave of religious activity, analogous in some respects to the spread of early Christianity, Buddhism, and Mohammedanism is passing over our American world. The common feature of these optimistic faiths is that they all tend to the suppression of what Mr. Horace Fletcher calls "fear thought." Fear thought he defines as "the self-suggestion of inferiority"; so that one may say that these systems all operate by the suggestion of power. And the power, small or great, comes in various shapes to the individual, power, as he will tell you, not to "mind" things that used to vex him, power to concentrate his mind, good cheer, good temper, in short, to put it mildly, a firmer, more elastic moral tone. The most genuinely saintly person I have ever known is a friend of mine now suffering from cancer of the breast. I do not assume to judge of the wisdom or unwisdom of her disobedience to the doctors, and I cite her here solely as an example of what ideas can do. Her ideas have kept her a practically well woman for months after she should have given up and gone to bed. They have annulled all pain and weakness and given her a cheerful active life, unusually beneficent to others to whom she has afforded help. How far the mind-cure movement is destined to extend its influence, or what intellectual modifications it may yet undergo, no one can foretell. Being a religious movement, it will certainly outstrip the previsions of its rationalist critics, such as we here may be supposed to be.

[...]

I have thus brought a pretty wide induction to bear upon my thesis, and it appears to hold good. The human individual lives usually far within his limits; he possesses powers of various sorts which he habitually fails to use. He energizes below his maximum, and he behaves below his optimum. In elementary faculty, in coordination, in power of inhibition and control, in every conceivable way, his life is contracted like the field of vision of a hysteric subject—but with less excuse, for the poor hysteric is diseased, while in the rest of us it is only an inveterate habit the habit of inferiority to our full self that is bad.

Expressed in this vague manner, everyone must admit my thesis to be

"The Energies of Men" (1907)

true. The terms have to remain vague; for though every man of woman born knows what is meant by such phrases as having a good vital tone, a high tide of spirits, an elastic temper, as living energetically, working easily, deciding firmly, and the like, we should all be put to our trumps if asked to explain in terms of scientific psychology just what such expressions mean. We can draw some child-like psychophysical diagrams, and that is all. In physics the conception of "energy" is perfectly defined. It is correlated with the conception of "work." But mental work and moral work, although we cannot live without talking about them, are terms as yet hardly analyzed, and doubtless mean several heterogeneous elementary things. Our muscular work is a voluminous physical quantity, but our ideas and volitions are minute forces of release, and by "work" here we mean the substitution of higher *kinds* for lower *kinds* of detent. Higher and lower here are qualitative terms, not translatable immediately into quantities, unless indeed they should prove to mean newer or older forms of cerebral organization, and unless newer should then prove to mean cortically more superficial, older, cortically more deep. Some anatomists, as you know, have pretended this; but it is obvious that the intuitive or popular idea of mental work, fundamental and absolutely indispensable as it is in our lives, possesses no degree whatever of scientific clearness to-day.

Here, then, is the first problem that emerges from our study. Can any one of us refine upon the conceptions of mental work and mental energy, so as later to be able to throw some definitely analytic light on what we mean by "having a more elastic moral tone," or by "using higher levels of power and will"? I imagine that we may have to wait long before progress in this direction is made. The problem is too homely; one doesn't see just how to get in the electric keys and revolving drums that alone make psychology scientific to-day.

My fellow-pragmatist in Florence, G. Papini, has adopted a new conception of philosophy. He calls it the *doctrine of action* in the widest sense, the study of all human powers and means (among which latter, truths of every kind whatsoever figure, of course, in the first rank). From this point of view philosophy is a Pragmatic, comprehending, as tributary departments of itself, the old disciplines of logic, metaphysic, physic, and ethic.

And here, after our first problem, two other problems burst upon our view. My belief that these two problems form a program of work well worthy of the attention of a body as learned and earnest as this audience, is, in fact, what has determined me to choose this subject, and to drag you through so many familiar facts during the hour that has sped.

The first of the two problems is *that of our powers*, the second *that of our means*. We ought somehow to get a topographic survey made of the limits of human power in every conceivable direction, something like an ophthalmologist's chart of the limits of the human field of vision; and we ought then to construct a methodical inventory of the paths of access, or keys, differing with the diverse types of individual, to the different kinds of power. This would be an absolutely concrete study, to be carried on by using historical and biographical material mainly. The limits of power must be limits that have been realized in actual persons, and the various ways of unlocking the reserves of power must have been exemplified in individual lives. Laboratory experimentation can play but a small part. Your psychologist's *Versuchsthier* [research animal], outside of hypnosis, can never be called on to tax his energies in ways as extreme as those which the emergencies of life will force on him.

So here is a program of concrete individual psychology, at which anyone in some measure may work. It is replete with interesting facts, and points to practical issues superior in importance to anything we know. I urge it therefore upon your consideration. In some shape we have all worked at it in a more or less blind and fragmentary way; yet before Papini mentioned it I had never thought of it, or heard it broached by anyone, in the generalized form of a program such as I now suggest, a program that might with proper care be made to cover the whole field of psychology, and might show us parts of it in a very fresh light.

It is just the generalizing of the problem that seems to me to make so strong an appeal. I hope that in some of you the conception may unlock unused reservoirs of investigating power.

11 Laura Nader's Third-Wave Energy Anthropology

THOMAS TURNBULL

ANTHROPOLOGISTS HAVE LONG STUDIED diverse societies and their modes of energy use.[1] As such, anthropology should be an excellent disciplinary lens through which alternative histories of human-energy relations can be discerned. That said, historians of this subfield, the anthropology of energy, often begin with North American scholar Leslie White (1900–1975), who in the 1940s proposed that culture, problematically singularised, evolved via increases in the availability and efficient use of energy.[2] In the 1970s, amid that decade's energy crisis, a second wave of studies led by a student of White's, Richard Adams (1924–2018), and Roy Rappaport (1926–1997) further exchanged anthropological nuance for scientific authority and made overly deterministic statements about the relationship between energy and society.[3] However, as that decade ended, a third wave kind of energy anthropology, developed by Laura Nader, criticised White and his aspirants for having offered an energetic "'grand theory' of social change" that linked changes in energy use "to social evolutionary ideas" which the discipline of anthropology had long sought to jettison.[4] Nader was also exceptional insofar as she sought to apply her anthropological theories to energy policy and did so from a perspective that was informed by her distinct positionality.

Mainstream anthropology was antithetical to grand explanatory theories. The roots of this idea had come from Franz Boas (1858–1942), the founder of cultural anthropology. Originally trained as a physicist, his doctoral research at Kiel University had attempted to explain why water assumed different colours at different depths. Seeking more data, he circumnavigated Baffin Island, now Canadian-Nunavut territory, aboard the research schooner *Germania* in 1883, a voyage during which he became interested in human variations in perceptual capacity. He took the indigenous Inuit as his subjects, learning Inuktitut, eating raw sea liver, and devoting himself to understanding their relation to the local environment.[5] The hardships and forms of reasoning he encountered among these people led him to reject any belief in cultural absolutes, hierarchies, or explanations that failed to consider science as culture. To think otherwise, he came to believe, served only to reveal one's own parochialism.[6] The Boasian school of anthropology he established affirmed cultural relativism.[7] Boas and his followers sought to distance themselves and their science from its past links to colonialism and scientific racism.[8]

As such, contemporary energy anthropologists walk a tightrope between White's explicitly anti-Boasian energeticism, with its attendant scientism and social hierarchisation, and a more conventional Boasian orientation toward particularism and the rejection of scientific notions of cultural superiority. At our current historical juncture, however misplaced the first wave's scientism might be, the second wave's approach risks an indefensible indifference to the many clear signs, in a dramatically warming world, that reforming current energy-society relations is of great importance to the survival of some of Earth's most vulnerable societies.[9]

Nader's third-wave energy anthropology, formulated in the '70s, offered a different approach. She sought to reintegrate Boasian cultural particularism as a means to provincialize statements made by those who claimed universally applicable relations between energy and society. She did so by affirming the very real material effects of energy use on the environment and for specific social relations. To reconceptualise energy in a particularistic way, Nader used a classic Boasian method, ethnography. With skills honed during fieldwork with Shia people in Lebanon and Zapotec in Mexico, she would later turn her observational acumen on another social group: those North American scientist-engineers, primarily male, entrusted with determining the scale and composition of the nation's energy system. In doing so, she documented and critiqued a resurgent form of energy determinism.[10] She instead argued the social consequences of energy use were not rigidly determinant, as White had suggested, but subject to shifting cultural norms. To believe otherwise, she believed, unhelpfully bulwarked, as the title to her *Physics Today* article stated, "barriers to thinking new about energy."[11]

In the spirit of this volume's attempt to encourage a more cosmopolitan understanding of energy history, here Nader's particularist view on relations between energy and society is situated within her wider biography and, to some degree, within the disciplinary currents of anthropology and the wider politics of the time.

Born in 1930, Nader grew up in Winsted, Connecticut, in an Eastern Orthodox Church family. Her father was a businessperson from Arsoûn, Syria, and her mother an Arabic teacher from Zahlé, Lebanon.[12] Nader recalled growing up with "talk of colonialism and imperialism" over dinner.[13] She went on to study at Wells College, in Aurora, New York, before going to Harvard to specialise in anthropology.

In 1957, having received a grant from the Mexican government, Nader left to carry out fieldwork in villages in the Rincón region of the Sierra Madre in Oaxaca, Mexico, a region on the cusp of oil-powered modernisation. The first cars had appeared around the time Nader began fieldwork, and a road to Oaxaca City neared completion. She considered the region an unusually harmonious one and analysed its court records in the hope of establishing how such peaceable relations had been sustained.[14]

On her return Nader moved from Harvard's women-only Radcliffe College to UC Berkeley, becoming faculty at twenty-nine. Sexism remained rife at this supposedly progressive institution. On her way to a meeting Nader was denied access to the faculty club because of her gender; undaunted, she climbed through a window.[15] As late as 1979, student activists revealed she and other women received salaries of 27,500 USD, half that of men.[16]

Increasingly disenfranchised by the prevailing legitimations of post-war American society, Berkeley's student body had embraced political radicalism in theory and practice, and in doing so they faced concerted attempts at their suppression.[17] Campus protests against the US-led war in Southeast Asia and colonial violence and domestic racism enforced in the latter stages of the Cold War were intermittently attacked by police armed with batons, CS-gas, and other projectiles; this violence served only to further reveal the militarised innards of US power.[18] Berkeley, she wrote, had become an "experimental station" for domestic counterinsurgency. During one particularly violently policed protest, she went on to note, an unknown noxious substance, which Nader termed "chicken shit" gas, had been dropped on protestors. California's governor at the time, soon-to-be president Ronald Reagan, later denied knowledge of its use and ingredients.[19]

Amid such turmoil, Nader advised students to "study up" and turn the ethnographer's tool-kit inward and upward toward the conventions and institutions of the US itself. What, she asked, could anthropology become if it studied "the colonizers rather than the colonized"?[20]

In the same period, it became evident that anthropology had become an accomplice of the US Cold War state. It has been estimated as many as five thousand academics had cooperated with the US Central Intelligence Agency (CIA) in the mid 1970s. Among these, Nader was appalled to discover the extent to which fellow anthropologists had become accessories to the maintenance of US hegemony.[21] She played a role in unveiling Project Camelot, a mid-sixties plot by the US Army that, it was hoped, could use anthropological insights to maintain control of or provoke insurgency in Latin America.[22] That said, her greatest ire was directed against US anthropologists working in Thailand in 1976 who had aided counterinsurgency operations during an anti-communist purge in which over 3,000 people had been killed by the army and during a later police attack on student protestors at Thammasat University, events later memorialised as the "October 6 Massacre." She defended the students who publicised these events, and faced censure from the American Anthropological Association for doing so.[23]

Statist anthropology was not new. As a graduate student, Nader's advisor Clyde Kluckhohn, a specialist on the Navajo people, was later revealed to have been moonlighting for the CIA. He had overseen politically motivated hiring and firing at the Harvard Russian Research Centre he led. His directorship, which begun in 1947, was strange given that he initially spoke no Russian nor knew much about Russian culture.[24] Nonetheless, Kluckhohn's *Mirror for Man* (1949) had called for a "repatriated anthropology," an applied science of reform in contrast to Nader's more radical call to study up.[25] Where Kluckhohn described a flawed but essentially well-meaning US, Nader took a more pointed view.

Mention needs to be made of anthropology's ongoing structural turn at this time. The work of Claude Lévi-Strauss, which appeared in the 1950s, advocated an anthropology that used ideas loosely derived from the natural sciences. In Lévi-Strauss's work physical notions such as symmetry and entropy assumed explanatory power while retaining a commitment to relativism; he argued that while the sciences suggested structures were invariant, their implications could be highly individualised.[26] His *Tristes Tropiques* (1955) described a number of indigenous cultures, predominantly in Brazil, whose traditional forms social organisation and structures of meaning were increasingly subject to processes of disorder. With its visions of irreversible degradation, dissipation, and cultural homogenisation, what Lévi-Strauss chose to call "entropology" was distinctly more pessimistic than White's energy-driven vision of social evolution.[27] Entropy-centred structural anthropology continued in the work of Lévi-Strauss's student Philippe Descola, who carefully recorded the calorie- and protein-filled diet of the Achuar people who lived alongside Ecuador's Pastaza River.[28] However, rather than looking to either law of thermodynamics for general explanatory principles, Nader's energy anthropology instead focused on the totems and taboos of those people who used the *most* energy: North Americans.

Before turning to this episode, we might consider the political stakes of Nader's epistemology. Anthropologist Hugh Gusterson has argued that her call for "studying up" encouraged a politically engaged approach to science ethnography around the time that anthropologist Bruno Latour and sociologist Steve Woolgar were observing scientists in action at San Diego's Salk Institute. Their concern with agency led to the influential yet less obviously political ethnography *Laboratory Life*.[29] Nader's approach was more pointed. For her, Polish anthropologist Bronisław Malinowski (1884–1942) had begun the sociological study of science with his 1925 essay "Magic, Science, and Religion," which ascribed a scientific mentality to the navigationally adept atoll-dwelling Trobriand Islanders and in turn a magical and a religious bent to staid Western scientists. Nader also credited another Pole, Ludwig Fleck (1896–1961), for having studied up before her in observing the social negotiation involved in laboratories treating syphilis in Lvov, Eastern Poland (now Ukraine), in the 1930s.[30] These progenitors seemingly imbued Nader with a theory of knowledge that saw a universal propensity toward scientific behaviour irrespective of a society's perceived developmental status, while in return she believed the scientific beliefs of so-called developed nations could be recast as far from rational.

The irrationality of one branch of US science would become directly apparent to Nader. In 1975 she was invited to participate in a major research project initiated by the Carter administration: Energy Modelling for an Uncertain Future carried out by the US National Academy of Science's Committee on Nuclear and Alternative Energy Systems (CONAES).[31] Nader recalled being the only woman amid three-hundred men on the project.[32] Nobel Prize–winning economist Tjalling Koopmans (1910–1985) had been tasked with overseeing six computer models that would forecast the consequences of changing patterns in energy use up to the year 2010, with special attention given to nuclear power as a potential alternative energy source.[33] The thrust of CONAES was that growth in energy demand was of central concern to the North American way of life. Without considerably more power or a marked reduction of growth in energy demand via conservation, national progress would falter.[34]

As background research, Nader headed a subproject looking at the concept of "lifestyle" and its relation to energy demand; this loose term described the patterns of consumption, location, and occupation that created overall rates of energy use, or so she was told.[35] Nader and her colleagues rejected a simplistic concept of lifestyle, dismissing it as a description rather than an explanation. Lifestyles were far more, they argued; they were enactments of a culture's values, ideational frames within which behaviour gained meaning.[36]

Beginning from this perspective, she worked with fellow anthropologist Stephen Beckerman to discern if energy use had an impact on "quality of life." They argued that asserting a positive (or negative) correlation between the two did an injustice to the complexity of their relation in practice.[37] In part, this was because, like lifestyle, she believed quality of life was a nebulous concept, something better suited to philosophical speculation or anthropological observation than quantification. Irrespective, as Antoine Missemer points out in chapter 4 of this volume, economists had long claimed increases in a gross national product (GNP) correlated with increased energy use. However, recent energy conservation initiatives appeared to refute this; comparative studies, particularly one contrasting "quality of life" in Switzerland and the USA, showed no clear correlation between per capita energy use and economic welfare. The outcome of this was that Nader and Beckerman found no "evidence that increasing energy use will increase the quality of American life."[38] It appeared the central assumption underlying the CONAES study, that the US needed ever more energy, was mistaken.

Between 1976 and 1977 Nader recalled occasionally working through the night to complete her part of the CONAES study, titled *Energy Choices in a Democratic Society* (1980).[39] In it, she argued that "energy is a social problem, not a technological one" and that the US had been misguidedly "founded on the principle of ever-expanding consumption."[40] The report sought to map out the social implications of a potential move away from energy expansionism by reducing total energy consumption and shifting demand toward renewables. Her study also described the threat of a nightmarish "large-scale bureaucratic centralization" if fast breeder nuclear reactors were deployed at the scale that the main CONAES report advised. Nader's supporting surveys of total solar power potential ("abundant" but "fraught with uncertainty") and opportunities for conserving energy ("significant" in number but with "both positive and negative impacts") suggested that an alternative energy future was entirely possible but it depended upon the "human component" of the system and the values which they, on aggregate, worked to realise.[41]

Nader's distinctly Boasian move was to point out that the cultural determinants of energy demand tend to be left implicit while scientific and economic rationales are foregrounded. In more pointed preparatory work, she expanded upon the misguided belief that technical specialists could be left to address the "human dimensions of energy issues," which she decried as being as ridiculous as having "a nuclear reactor built by a social psychologist."[42]

The problem was a persistent Western-centric belief that societies developed in step with ever increased levels of energy consumption (Chatterjee, chapter 5, and Introduction, this volume). Alongside the usual suspects, Ostwald, Soddy, and Lotka, she considered archaeologists Grahame Clark (1907–1955) and V. Gordon Childe (1892–1957) as antecedent advocates of energy-driven social evolutionism, a movement which appeared resurgent in the '70s.[43] This time, she argued, figures such as ecologist Howard Odum (1924–2002) had begun advocating a new energy-determined view of society. For Odum and his information theorist co-author Richard Pinkerton, energy was a fundamental input amid an emergent complex of feedbacks.[44] While less mechanistic than earlier proposals, this reformulation of energy/society relations led to a familiar conclusion: the more elaborate a system's energetic feedback loops, the slower overall entropy increased and the better quality of life that could be achieved.[45]

Following the publication of *Energy Choices*, Nader was invited to give a talk at the MITRE Corporation, an engineering think tank. She was taken by limousine to defend her findings to a potentially hostile audience.[46] In response she gave a humor-laced talk which was later published in 1981 in an edited form in *Physics Today*, the in-house journal of the American Physical Society, which is presented here. Playing with anthropological conceits, she spoke of how over the course of the CONAES project, she observed how the energy community was guided by an overarching ideology: increased energy supply was its totem. Moreover, though viable alternatives to fossil fuels and nuclear power existed, they were often marginalised by self-declared policy experts and engineers due to their perceived lack of scientific interest.[47] Her talk sought to explain how these federally funded experts, a distinct cultural group, engaged in value-laden practices.[48] For instance, these experts considered low-energy lifestyles, of the kind that were typical in large parts of the world, as unthinkable.

She also described the absence of gender diversity as a clear problem. Like any single-sex group, she believed this composition encouraged competition, and—given the dominance of males—gave rise to irrationally paternalistic ideas about certain technologies having been "fathered." This led to the observation that these experts had a curious tendency to gender their objects of inquiry. Energy conservation, with its hint of household probity, was considered feminine but unsexy. While nuclear power, dangerous, prolific, and potentially fast breeding, was cast in a masculine role. Solar, given its unloved nature, was "an orphan child," something of low technical interest and of little concern to elite scientists.[49] While never claiming to be a conventional feminist, Nader's positionality no doubt helped her recognise how this male-dominated social group enforced a number of irrational and gendered norms which limited the scope of possible interventions in the energy system.[50] Given this propensity, as she later put it, repeated assertions of the objectivity of their analyses did "little else than conceal the scientist's highly subjective position."[51]

Nader later recalled how, as an index of its success, her admittedly provocative article encouraged heated letters from scientists and engineers. One national laboratory scientist broadly agreed with her, arguing it was the sciences' professionalisation that had led to such corrosive "group think." He went on to argue that the people she had met "were not concerned with the energy problem so much as getting a piece of action"; while a physics professor, as if invoking Malinowski, bemoaned the veneration of ever-increasing energy supply as "the American Religion," something people unthinkingly "accept without subjecting it to analytical scrutiny." Like all religions, its self-serving "high priests don't want to hear heresy."[52] Another respondent remained faithful, arguing that in contrast to those who work with people (presumably he meant anthropologists and other social scientists), "those who work with numbers and objects almost without exception have improved the well-being of everyone."[53] Nader later noted how the responses suggested scientists were far from rationality led. Riled scientists, like Trobriand Islanders, or anyone else, exhibited "reason and desire intermingled."[54]

Nader and Beckerman hoped energy anthropologists could become indispensable to diverse civic groups concerned with "the energy problem," from Navajo tribal councils to Appalachian coal miners.[55] However, the 1980s were marked by a widespread decline in concerns about energy as oil prices returned to pre-crisis levels. Low prices encouraged US companies to invest in cheaper Persian Gulf oil rather than to increase domestic production, a move enabled by further US militarisation of the Middle East.[56] In doing so, US hegemony became more entwined with the oil that geological circumstances had left beneath Gulf States.[57] At home, Reagan's commitment to "New Federalism" meant budgets for alternative energy and energy conservation research were cut to the nub, while vast subsidies continued to be doled out to nuclear industry.[58] Nader's view on later US-led invasions in the Middle East led her to reaffirm a finding from her Oaxacan fieldwork: "harmony coerced is freedom denied."[59]

In light of the growing evidence and evident impact of climate change, energy became a resurgent concern in the 2000s. Nader reviewed a number of published energy ethnographies with rightful satisfaction. She conveyed how anthropologists had linked Exxon's operation to supernatural belief systems in French Equatorial Africa, and how Norway's successful Statoil concern overturned a blanket belief in the "resource curse."[60] However, she also complained of contemporary anthropology's growing "fogging process," its growing "delight in the abstract" rather than in empiricism.[61] Fearing accusations of extractivism, anthropologists have increasingly turned to theorization rather than seeking application. Yet in these same years, the effect of more parts of the world committing to energy-expansionist lifestyles became dangerously apparent.

Over Nader's career, roughly from 1950 to today, the global consumption of energy increased by around ~22 zetajoules. This is more energy than we used in the preceding 11,700 years (~14.6 zetajoules).[62] As a result, humankind has become an Earth-altering geological force, a situation that Nader took as a means to validate her claim that cultural changes ultimately determine patterns of energy use, rather than energy being determinant of cultural change.[63] Today, the scale of human impacts on Earth makes some kind of causal relation between energy and society appear irrefutable. However, those offering overly simplified correlations and presumed mechanisms of determination risk falling back on old theories of energy determinism. Thankfully, Nader's work, in pointing to the need for culturally specific and value-centred analyses of society-energy relations, offers an approach that helpfully specifies rather grandly abstracts such exchanges. Most importantly, in relativising energy-society relations, as Boas had done for socio-evolutionism, her work suggests the possibility of forging such relations anew.

LAURA NADER
"Barriers to Thinking New about Energy" (1981)*

Anthropologists study the past and the present; we don't normally study societies that don't exist, let along invent them. Until about five years ago, I specialized in studying legal systems cross-culturally. Now I also look at questions about energy, science and expertise, the work place of scientists,

---

*Laura Nader, "Barriers to Thinking New about Energy," *Physics Today* 34, no. 2 (1981): 9–104. Based on an address delivered at the MITRE Corporation, McLean, Virginia, on 6 November 1979. Reprinted with the permission of AIP Publishing and the author.

"Barriers to Thinking New about Energy" (1981)

and the freedom of science. How did an anthropologist get involved in energy issues?

A few years ago, some people at NASA asked me to attend a conference in Monterey, California, on portable energy systems. My immediate response was that they had the wrong Nader. The man at the other end of the line said, "You are the anthropologist? We want an anthropologist at this conference." So I went down to Monterey for five extraordinary days with different groups of professionals, mainly scientists and engineers. We talked about different "scenarios" for the future. (I hadn't even heard the word "scenario" at that time.) We were to think freely about these different scenarios for the future, but it became quite clear that there were already boundaries around those scenarios. You were to think freely—within those boundaries. When you went beyond them, someone would tell you, "You're off the track." Finally I told one fellow that we didn't know where the track was; that was why we were there.

At the closing session, I told these people what I thought, as an anthropologist, listening to them. I sketched why I had gone into science—for anthropology is a science as well as a humanity. One reason why I'd gone into science was that I was curious; I believed that scientists encouraged curiosity. In Monterey I had found that, in fact, curiosity and the freedom to roam mentally were curbed among physical scientists and engineers.

The group itself was very limited: all but three were men; all of us were white; and aside from two social scientists and two lawyers, the rest were engineers and physicists.

The most striking observation was the number of taboos. Solar was never mentioned by anybody other than myself, literally not mentioned. The possibility of dropping nuclear power as a future alternative wasn't even discussed. The social and political consequences of nuclear power were not discussed. Nobody used the word "safety." These were all taboo areas. The fact that we were making decisions that closed off options to the next generations was considered irrelevant.

None of these and other central issues were talked about. Every time I raised them people would say things like, "You remind me of my son." That gave hope. They were at least raising their children right.

Laura Nader

After my talk an engineer smiled at me and said, "Professor Nader, I would like to explain why we don't talk about safety; we don't talk about safety because it's built into the design." As an anthropologist I found that statement interesting enough to write down.

A man from Texas said, "I've been on a lot of big work projects all my life. Nuclear's no different. Anytime you're building something, you lose some people. It's just a big crap game, and this is the biggest crap game of them all. You toss the dice and hope you hit seven."

My eyes were opened further at a lecture about breeder reactors at the Lawrence Berkeley Laboratory that I attended with my husband. The same sort of group was there—mostly white, mostly male, and mostly scientists and engineers. The man who introduced the speakers said, "Since breeder reactors are the way we're going to go, we've brought two people from Atomics International to discuss the question."

In the first place, who says that breeder reactors are the way we're going to go? There was no discussion of that question either before or after the talk.

The first question came from a young man in the front row, who said, "I find it incredible that you've talked for a whole hour about the breeder reactor and never raised the question of public safety." I said to my husband, "That man does not work in this laboratory." (He didn't, either. He was John Holdren, a professor on the campus.) Several questions followed, but the only ones about public safety came from young graduate students. Not a single worker at the laboratory asked about public safety.

For the first time, I began to question how the work organization affects how scientists and engineers think. There are certain pressures, at that laboratory and others like it, that encourage people to think similarly—that, in fact, punish deviant thinking. In science, new ideas come from oddball thinking and freedom of expression. After that, I kept my eye on the work place of the scientist and engineer.

I was soon involved in the CONAES (Committee on Nuclear and Alternative Energy Systems) project at the National Academy of Sciences. The CONAES was divided into the Synthesis Panel, the Supply Panel, the Demand Panel and the Risk Panel. I was a resource chairperson for the Synthesis Panel. My group was to describe what life would be like in the year

"Barriers to Thinking New about Energy" (1981)

2010 under different levels of energy expenditure. We were going to go from 70 quads to 70 quads, from 70 to 110 quads, from 70 to 175 quads and so forth on to 2010.

I was intrigued by how people were working on the project. In the first place, I'd never done any work with the future. As I've said, anthropologists study the past and the present; we don't study societies that don't exist, nor do we invent them. I soon learned that our humility was probably misplaced in this project, because economists don't mind inventing all kinds of societies. When what they invent often happens, invention becomes self-fulfilling prophecy, much to many people's horror.

I noticed the earlier patterns in this group: a good deal of standardized thinking; lack of respect for diversity; absolute taboo on the word "solar." Their memos discussed nuclear, coal and non-nuclear. Non-nuclear was solar.

I asked the co-chairman, "How come nobody ever uses the word 'solar' around here? I've been on board six months and nobody's used the word 'solar.'" He looked at me, rather surprised. "I don't know. Solar's been an orphan child." Somebody else piped up. "Solar? Solar's not very intellectually challenging." Somebody else said, "What's solar? A bunch of mirrors."

Some things said off the tops of people's heads have much deeper meaning, as any social scientist could understand. The first observation was, "It's an orphan child." The president of the American Chemical Society in 1900 predicted that the US would be running on solar by the 1970s. When did it become an orphan child? Did it have anything to do with World War II, the nuclear developments, militaristic interests, and so forth? What are the reasons for that? One could write a paper just on that observation.

The other observation: "Solar's not very intellectually challenging." What is intellectually challenging to these people? They seem to relish something complicated, hazardous, difficult and risky, something that requires high technology and big money. They seem to have a real attraction to that sort of thing. I'm not a psychologist, but I came as close to psychologizing during the CONAES study as I ever have.

I had on my resource team two physicists, a computer technologist, one sociologist, an economist, an engineer—not many social scientists. They were a diverse group, willing to experiment with different futures. We started out with the idea that energy demand was not going to expand by 2010; it would stay at 70 quads. The challenge was, could you go from 70 to 70 without changing amenities, in such a way that people wouldn't be disturbed or disrupted in their lives. The value system, essentially, wouldn't change.

One economist reported that it was impossible to go from 70 to 70. We asked him, "Have you concluded that it's impossible because you're using growth models?"

What do people think is possible? Why are people so tightly constrained? Perceptions of the past, the future, and the present are intimately tied to what one feels is possible, and what one is optimistic or pessimistic about.

Our 70-to-70 scenario is fairly easy to carry out, with little disruption in people's lives. Essentially what we focused on was technical efficiency. Cars get more miles to the gallon, refrigerators give the same service but use less electricity. We had gimmicks that people could use to turn on and off their gas or oil in the house; lots of little things that added to a fair amount of saving with very little change.

Going from 70 to 70 was easy, because there's so much waste in the system, so we decided not to go from 70 to 110 as directed, but down to 50 instead. We got some interesting reaction to that scenario. Who ever heard of going down without going backward?

Many people misunderstand the direction of change and the ways societies change. In the 50-quad scenario, most of the responses to problems are bottom up. The reason that people can't understand that scenario is because professionals in this country tend to think top down. Even where this does not happen, where there is ample evidence of the direction of bottom-up change, people in power believe that change comes from the top down.

Major changes in demographic factors, for example, are not top-down changes. They are individual decisions made in households all around the country. The invention of the car was a dramatic change that started as a small industry and diffused.

Our 50-quad society was a bottom-up change scenario. We made it that way on purpose. It was not a utopia; it was an exercise to make people think a little differently. We may not even want to live in such a society but we

"Barriers to Thinking New about Energy" (1981)

wanted to juggle people's thinking about what is possible. We wanted to point out, first, that it is possible to have a high-technology, low energy society, and second, that change can come from more than one direction.

While we were working, no matter what we sent to Washington, we would be asked for more tables and less prose. We finally got an exasperated note that said, "More tables, less prose. These guys don't read." We know there's a literacy problem among the young, but less recognized is another serious problem in this country: managers and planners do not read and they do not write. They hire people to do it for them.

For people who want it all in tables, I ask: "How do you talk about freedom in tables? How do you talk about democracy in a table? How do you talk about most of the things we care about in a table?" We compromised: we used both prose and tables. It's probably one of the few reports that can be read by the tax-paying public.

The CONAES project was the hardest field work I'd ever done. I've worked among Indians in southern Mexico, Shiite Moslems in southern Lebanon, and in a variety of places in this country. This was, by far, the most difficult work. I think it's important to understand why.

People in technical areas work with objects or with numbers. They don't work with human beings. When you have people who never work with humans thinking about our future, sometimes humans are treated like objects.

The kinds of statements that I heard at CONAES might not seem strange to you, but I want to repeat them because they are very strange. In the Risk Panel, a well-known risk specialist, in reporting his conclusions, said, "Fifty thousand people die in car accidents every year. We know how to build cars so that doesn't happen. There are X number of people who die in dam breaks, from household accidents, and from various other accidents. We know how to prevent those deaths. If we prevented those deaths, then we could afford to have a nuclear disaster." I understood why, when I came home from this work, I would head for the shower before I would greet my children. That kind of thinking is truly polluting.

The absence of diversity leads to serious problems. To start with, male professionals in this country are very macho. I don't say that facetiously. I really did not envy the men of that committee. I'd never watched men operate in groups like that, because although anthropology is still predominantly male, it's very heavily coed.

Something happens to same-sex groups, they vie with one another. You can see this in the non-human primate literature as well. Same-sex groups are very competitive. In this case, big is better. Hazardous is interesting and intriguing. I learned recently in Los Angeles that conservation is considered feminine. Isn't that interesting? That must make nuclear a very masculine endeavor.

After the CONAES study, I worked on yet another study. The DOE gave some money to University of California groups to do a study of what's called soft energy paths (Amory Lovins's term) in California. Can we go to soft energy paths, soft meaning non-nuclear, meaning decentralized, meaning solar usually? We began a similar kind of futures exercise that we were doing with CONAES.

Again and again I saw people's methodologies getting in the way. They were using old methodologies that were appropriate to other problems like growth modeling, to see what it was going to be like with less in the year 2010. They were tripping all over their methods and coming out with fancy computer statements that had little credibility. Shamans would evoke more confidence.

As an anthropologist I find this wedding to numbers fascinating. The belief is so strong, it's like numerology, the belief that numbers in themselves are useful. Numbers are useful, of course, but not in and of themselves. In a controversy, when one side gets a numbers man and the other side gets another numbers man and the two sides fight with numbers, and count coup, numbers have lost their utility. Both studies suffered from this belief that numbers in and of themselves add strength to an analysis of the future.

At one point in the CONAES study, I visited a breeder-reactor group meeting at EPRI. At that time, I worried about losing my objectivity, so I took a professional linguist with me. We heard things like this: Jack says to Bill, "Bill I like your numbers, they agreed with mine." Bill beams and says, "How about Jim? Has Jim generated any numbers yet?" Jack says, "No. Why

"Barriers to Thinking New about Energy" (1981)

don't you send your numbers over to him before he gets his ego involved with generating his own." I couldn't have made up these lines.

In a way I'm glad that there's humor in this, because there is so little humor in the whole energy question. But I don't want the humor to mask the importance of such observations or their consequences.

In the California soft-paths study, we took a look at two kinds of change: top down and bottom up. We looked at the mandated solar code; that was top down. We looked at the possibilities of distributed energy, which was bottom up. People could create their own wind and electrical systems and then feed it into the grid.

I found myself looking at work patterns again. The code we looked at dealt with encouraging solar energy use, natural gas, insulation, glazing and so forth. The people who wrote the code, I think, were inadequately aware of the human component. However, it became quite clear, in talking to the different people involved with this code, that certain people will determine if that kind of a mandate works or doesn't work.

We interviewed a wide range of people from different interest groups: bankers, contractors, architects, building inspectors, lawyers and realtors. Each type of worker belonged to a particular subculture of work, with an organization and value all to itself. They each had almost unique ways of looking at building codes. It was extremely difficult for anybody we interviewed, except members of the general public, to see the whole picture. Everybody saw the picture that impinged on their individual self-interest.

As a group, architects react critically to codes. They see them as a hindrance to their creativity, particularly if they mean more paperwork. Building inspectors, who are already overworked and understaffed, worry about even more work. Realtors are doing well as they are, so why should they endorse codes? Government bureaucrats are straitjacketed by the organization, and only able to work their mandates, even if the solution to their problem might lie outside that mandate, as narrowly denned. The utility planners had a similar difficulty. They found it hard to think of utilities as generators or sellers, rather than as buyers, of energy.

None of our material on workers is new data, but look at the meaning of such data for transitions we're coming into now. Realize that these very

Laura Nader

difficult times are not different because of any natural resource scarcity but because of such facts as self-interested workers who aren't rewarded for deviating or changing. They're rewarded for doing things the way they've always done them.

If I were an anthropologist from New Guinea, observing the energy efforts of the past several years, I would note a wide gap between what leadership says and what it does in this country. I would note that the government had no serious interest in solar. All the solar conferences the government is sponsoring, I would see as rituals of reconciliation. In the absence of true innovation and change, we have one conference after another. Because of the way American leaders are handling the problem, I may theorize that the society is having a nervous breakdown instead of an energy crisis. I would be struck by the presence of solutions in the absence of will. One conservation researcher at the Lawrence Lab cut electricity use over 40 percent in a major building without anybody noticing. Yet most of our federal structures have not begun serious conservation.

Conservation isn't sexy. It's not hazardous; it's not risky: it's obvious. We have gotten to the point in our society where we can no longer entertain obvious solutions.

This is where anthropologists come in. The coming era will require practical, general, earthy types of thinkers who understand problems and conflicting value systems. We need people who can look at mundane and straightforward problems, people who will not choose complicated solutions when simple ones are available.

The energy problem is not a technological problem. It's a social problem. We must build technologies that recognize human frailty. If there's one thing that social science has documented, it's that people make mistakes. They're going to continue to make mistakes. Build that into the technology, and accept and reject technologies on that basis.

You must look at the concept of progress and decide whether, in fact, simple progress is what we have now. Anything we do we label as progress. We must decide whether progress as a concept should be reserved for something that improves the quality of life.

The toughest problem will be getting professionals to look inside them-

"Barriers to Thinking New about Energy" (1981)

selves, to see what their mind-set problems are. What is it about my anthropological training that makes me see things in a certain way? What is it about your technical training that makes you see certain things and not others? No one is comfortable exploring these questions about themselves, but it's part of the job that has to be done.

12   The Master Resource

Energy, Inter-Planetary Capitalism, and Neoliberal Cornucopianism

TROY VETTESE

"**IF I WERE A GAMBLER,** I would take even money that England will not exist in the year 2000," Paul Ehrlich prophesized as he boarded an airplane at Heathrow in 1969.[1] The controversial biologist, the *New Scientist* breathlessly reported, then "flew off into the night, leaving behind yards of exciting headlines."[2] Predictions of mass starvation, energy scarcity, and civilizational collapse abounded during the "Malthusian moment" of the late 1960s and early 1970s, when scientists combined grim soothsaying with rockstar popularity.[3] Ehrlich became better known for *The Population Bomb* (1968) and his appearances on the *Tonight Show* than for his scholarship on butterfly and plant co-evolution.[4] The neo-Malthusians differed amongst themselves in terms of rhetoric and willingness to countenance violence, but they all echoed the eponymous eighteenth-century vicar's bleak geometry of population growth exceeding agricultural production. The intersection of these curves would precipitate what Thomas Malthus called the "checks" of war, famine, and disease.[5] Ehrlich's own best-selling polemic was but one of a burgeoning neo-Malthusian corpus that included Paul and William Paddock's *Famine 1975!* (1967), Garrett Hardin's "The Tragedy of the Commons" (1968) and "Lifeboat Ethics" (1974), the *Ecologist*'s "Blueprint for Survival" (1972), and Donella Meadows *Limits to Growth* (1972). Neo-Malthusian pessimism stemmed from the trends of decolonization, financial turmoil, environmental deterioration, and excessive consumerism, all of which undermined the faith of lay people and experts alike in sturdiness of the postwar order,

Neo-Malthusians, however, were not the only political upstarts benefitting from mid-century malaise—the neoliberals would soon begin their own rapid ascent. This latter group were distinguished by their belief that the market was not merely a site of exchange, but one of knowledge production superior to any other institution, such as state planning or even science.[6] The neo-Malthusian-neoliberal rivalry was personified by the antagonism between our airport prophet Ehrlich and the "doom-slayer" Julian Simon.[7] He and other neoliberals realized that they needed to thwart the neo-Malthusians' rise by calming the public's anxieties over energy and resource scarcity if they were to triumph during this period of transition. Simon, a hitherto obscure business professor at the University of Illinois at Urbana-Champaign, fired the opening salvo in 1980 by publishing "Resources, Population, Environment: An Oversupply of False Bad News" in *Science* (which is reproduced at the end of this chapter). He criticized neo-Malthusian assumptions while sketching a novel cornucopian framework, which would later be called "resourceship" by his followers. *In nuce,* resourceship was predicated on the assumption that the market will always overcome problems posed by too few resources or too many people. Simon's essay provoked Ehrlich to begin a short-tempered correspondence and a long-term bet on the prices of copper, chrome, nickel, tin, and tungsten.[8] The two agreed that the market would register scarcity in terms of higher prices, and thus when their ten-year wager was called in 1990 and these metals cost half as much as they did in 1980, Ehrlich lost and posted Simon a cheque without a note of congratulations.[9] To be sure, the neo-Malthusians had already fallen significantly since the heady days of the late 1960s, but Simon's victory nonetheless signalled the neoliberals' growing influence in environmental affairs.

Resourceship is best understood through its three interlinked concepts of the "ultimate resource," the "master resource," and resource "services" which Simon presented in an inchoate form in "Resources, Population, Environment" and refined in succeeding years. First, human capital is the "ultimate resource" created by "skilled, spirited, and hopeful people who will exert their wills and imaginations for their own benefit, and so, inevitably, for the benefit of us all."[10] Rapid population growth would not necessarily be a cause for concern because human capital would be increased simultaneously. Second, energy was the "master resource" because "energy is the key constraint on the availability of all other resources."[11] Cheap energy allows the extraction of ever more diffuse or difficult-to-access resources, and it aids the synthesis of scarce commodities. Plentiful energy can make up for scarce resources in other ways, such as making large-scale water desalination and transport possible to irrigate otherwise dry regions.[12] Third, Simon stressed that the "services" derived from resources matter, rather than the resources themselves. He offered the example of copper wiring for telephone lines being replaced by satellites; thus the service of communication remained cheap even as copper became scarce. Resourceship emerged from linking these three concepts: the "ultimate resource" of human capital is used to discover novel means of employing the "master resource" of energy, which in turn creates or explores other resources that keep "services" cheap.

The purpose of this essay is not to detail the history of Erlich and Simon's bet, but rather to examine resourceship's intellectual context and its relationship to energy. While resourceship was originally a fringe theory far removed from the neoclassical mainstream of economics, in the ensuing decades it has influenced scholarly and lay environmental debates. Despite its ubiquity, resourceship has been largely ignored by energy and environmental historians.[13] Energy—especially nonconventional fossil fuels—plays a crucial role in resourceship because Simon believed enough cheap energy would allow capitalism to circumvent seemingly impenetrable barriers to endless growth. As his *Science* essay was published just a few months after oil prices peaked in April 1980, this cornucopian argument was not an easy one to make. Once the reserve of energy resources was expanded to include nonconventional fossil fuels and imagined interplanetary deposits, then the contemporary energy crisis appeared as a temporary inconvenience to a glorious future where humanity in its teeming trillions colonized the galaxy all watched over by markets of loving grace.

Simon's intellectual biography sheds some light on how he came to devise resourceship. Like Ehrlich, he was born into a family of Jewish immigrants in 1932, and the two men even grew up nearby in New Jersey. At different points of their education, Ehrlich and Simon both studied at University of

Chicago, though they pursued their seemingly incongruous disciplines of lepidopterology and advertising. Their biographical trajectories converged again in the early 1960s when Simon tired of studying mail-order marketing and turned to the voguish topic of overpopulation. Initially, he was gripped by the prevailing neo-Malthusian panic and offered his services as an advertising expert to family planning organizations.[14] Yet, Simon became more sceptical the deeper he delved into economics and demography.

The three main influences on Simon were institutionalist economics, revisionist economic history, and neoliberalism. Much of the heart of resourceship is taken from Erich Zimmermann, an interwar institutionalist, who argued counterintuitively that resources were not *discovered* but *created* by human ingenuity.[15] This insight was inspired by contemporary breakthroughs achieved by chemists in his homeland of Germany ("the classical land of ersatz") including synthetic petroleum, rubber, and fertilizer.[16] Armed with enough energy and pressure, humanity could wrest a much greater bounty from nature. Zimmermann inspired younger economists, such as Harold Barnett and Chandler Morse, whose monograph *Scarcity and Growth* (1963) discovered that natural resource prices have been in decline since the nineteenth century.[17] Their finding suggested that natural resources were becoming *more* plentiful despite decades of extraction. Concurrent to the efforts of Barnett and Morse, revisionist economic historians including Ester Boserup and Richard Easterlin argued that there was no correlation between economic decline and population growth. Simon Kuznets even ventured that "more population means more creators and producers."[18] Yet, even these fairly cornucopian economists conceded that growth would eventually stop. For example, Zimmermann warned that fossil-fuelled civilisation was "a passing phenomenon in human history," which would necessarily be succeeded by a less dynamic economy predicated on renewable energy.[19] This caution became common sense after the first oil crisis hit the global North in 1973. Just weeks after the Organization of Arab Petroleum Exporting Countries (OAPEC) refused to sell oil to Israel's European and North American allies, Robert Solow warned the members of the American Economic Association that in the very plausible event a limitless "backstop" source of energy was not found then humanity would face a "cold winter."[20]

Simon intervened in this debate by marrying the insights from revisionist economists like Kuznets with his newly acquired neoliberal faith in the market. This faith was absolute—he compared it to remaining confident mid-air after an "energy jump" from the Eiffel Tower. Instead of hitting the ground, he predicted that humanity was more likely "in a rocket that [. . .] will take off sometime soon."[21] Simon's zeal was unusual even amongst neoliberals. The patriarch of the neoliberal movement, Friedrich Hayek, admitted in 1960 that the "consumption of irreplaceable resources rests on an act of faith."[22] He believed that the transition from coal to petroleum validated this faith because the "warning of the conservationists" turned out to be unnecessarily cautious.[23] Yet, Hayek's own faith was tested by the rise of the neo-Malthusian movement. In the year after the first oil shock, a rattled Hayek attacked *The Limits to Growth* for its "far-reaching claims [. . .] made on behalf of a more scientific direction of all human activities," but failed to offer a robust defence of the market.[24] Hayek himself recognized the importance of Simon's breakthrough when he wrote a "fan letter" in 1980 because reading Simon's work had been "too exciting an experience."[25]

While Simon delineated resourceship in greater detail in his other works, such as *The Ultimate Resource* (1981), "Resources, Population, Environment" matters because it was the first time he presented resourceship to a broad audience and thus shifted mid-career from an apolitical business professor to a neoliberal bruiser. His article began by questioning the *grande peur* of the 1960s and 1970s—that the global South first then the rich North would be wracked by massive famines. He cited an article in *Newsweek* that reported an estimated 100,000 people had died in the Sahel during the 1968–1973 drought, but dismissed this figure for its "flimsy" evidence.[26] There had been no famine, Simon countered, just higher-than-average child mortality.[27] His point was that if "bad news about population growth, natural resources, and the environment" was based on "contradictory evidence," then the neo-Malthusian consensus "ranging from B. F. Skinner to Solzhenitzyn" was unjustified.[28] Simon then set up and knocked down various neo-Malthusian "myths," such as urban sprawl and the finitude of natural resources, to clear the stage for resourceship as his riposte.

In "Resources, Population, Environment," one can already detect resourceship's three main concepts even if Simon did not use those terms exactly. For instance, instead of saying the "ultimate resource," he discussed how the "ultimate constraint" on producing "unlimited raw materials" was "knowledge," and therefore more people meant more knowledge.[29] He hinted at energy's special property as the "master resource" when he declared "copper can be made from other metals."[30] Moreover, energy was not only becoming "more plentiful" but nigh infinite as "the earth does not bound the quantity available to us."[31] As the master concept in Simon's framework, energy's crucial role in resourceship was to make neoliberal cornucopianism seem plausible at a moment marred by scarcity, inflation, and economic decline. Simon used the term *services* in the essay, explaining that "as consumers we are interested in the services we get from the raw materials rather than the raw materials themselves" and used the aforementioned example of copper wire and communication satellites.[32] In this way, he linked arguments based on demography, chemistry, and economics by claiming that more people would allow humanity to "find new lodes, invent better production methods, and discover new substitutes, the ultimate constraint upon our capacity to enjoy unlimited raw materials at acceptable prices is knowledge."[33]

As a true neoliberal, Simon deferred to the market about which source of energy should power capitalism's endless growth. This is one way we can see how neoliberal cornucopianism differed from contemporary iterations. For example, New Deal Prometheans like Alvin Weinberg and Philip Hammond believed that a gigantic government-led effort to build 32,000 fast-breeder reactors would expand "the limits to population set solely by limits to our energy production" to be "considerably larger than 20 billion."[34] By contrast, Simon was neither antagonistic nor enthusiastic regarding nuclear power. He was bothered more by the fact that nuclear power was not as cheap as its proponents claimed than by the danger of radiation or a meltdown.[35] After all, he cared about the "services" various energy systems provided, not the systems themselves. If there was not enough petroleum, then it did not matter if nuclear or solar power substituted instead. Notably, Simon's "all of the above" approach to energy policy (to quote a twenty-first-century president) included nonconventional fossil fuels.

If conventional fossil fuels are how one normally imagines petroleum extraction—say, with a "nodding donkey" drawing oil in twentieth-century California or Saudi Arabia—then nonconventional fossil fuels differed in terms of extraction (e.g., deep-sea rigs, "fracking," etc.) or the resource itself (e.g., kerogen or bitumen rather than petroleum). Simon saw the potential for the tar sands and other nonconventional fuels to act as Solow's "backstops" at a time when such industries were tiny. When Simon was writing in the early 1980s, only three countries had formerly had substantial nonconventional fossil fuel industries: Nazi Germany and Apartheid South Africa—which both sought to protect themselves from blockade and embargo by relying on coal-to-oil synthesis—as well as Canada, with its small bitumen industry in northern Alberta. To put Simon's ideas into context, there was a renewed interest in nonconventional in the US during the 1970s, as politicians debated whether the government should foster the country's kerogen industry to reduce oil imports from the Middle East. While neoliberal standard-bearer Milton Friedman dismissed the proposal as a "boondoggle," Simon believed that eventually a nonconventional industry could be conjured into being by market forces.[36]

Nonconventional fossil fuels further revealed the nuances of Simon's neoliberal epistemology. He sought to blur the boundary between conventional and nonconventional fossil fuels because the estimates of deposits would become "increasingly loose, and hence less 'finite' and 'limited.'"[37] In this way, he was attacking the foundational assumption of neo-Malthusianism that the world was composed of discrete objects that could be measured by experts. Simon believed that experts could not determine "the future quantities of a natural resource [which] cannot be calculated even in principle" because market-led discoveries of resource deposits and substitutes could not be known beforehand.[38] It was impossible to predict *which* resources would be used (such as nonconventionalists), as well as *where* these resources would be extracted. Simon expanded the latter category to even diffuse elements in seawater as well as the solar power from "other suns."[39] These epistemological considerations set neoliberals apart not only from the neo-Malthusians but also the mainstream neoclassical economists like Solow, whose models depended on the total resource base being known a priori. Rather than restraining the market and population growth within limits as the neo-Malthusians demanded, neoliberals believed that such restrictions would impair the market's epistemic functions when they were most needed—a risk that justified Simon's seemingly reckless faith.

Given the importance of energy as the "master resource" in Simon's framework, it was hardly surprising that it was precisely on this point that Ehrlich, his wife Anne, and their co-authors Paul Holdren and John Harte focused their critique. The three penned an acerbic letter to the editors at *Science* in response to Simon's essay. While they concurred that Simon's concept of services was not necessarily wrong, substitution was not a costless process. As easily accessible reserves were used up, "the delivery of services by materials-intensive or energy-intensive" resources necessitate higher prices. Therefore, alternatives like nonconventional fossil fuels would "ameliorate, not eliminate, the costs of the scarcity of oil." While Simon had breezily invoked the possibility of "irrigating deserts" to increase arable land, the Ehrlichs and Holdren dismissed desalinization as too expensive to serve as a substitute for rivers and rainfall.[40] The three neo-Malthusians could not maintain a veneer of polite disagreement when they discussed Simon's insouciant claim that "copper can be made from other metals."[41] "Perhaps Simon here has in mind the technique of elemental transformation by bombardment with subatomic particles in accelerators," they speculated. Even then, it would be a "gargantuan feat" to produce even "microgram quantities" of copper.[42] The Ehrlichs, Holdren, and Harte warned that resource substitution predicated on fossil-fuel extraction and deforestation would engender ever greater side effects by releasing a "pulse of carbon dioxide [. . .] to alter global climate in a way that undermines food production to an unprecedented degree." A deteriorating biosphere would offer humanity fewer ecosystem "services" that could not be easily substituted because of "the intricacy and the immensity of these processes." Theirs was a damning critique of resourceship that threatened to strangle the concept in its cradle.

Ehrlich's weakness was his temper. Simon was blatantly trying to bait Ehrlich to piggyback on his fame, so he could shift public and scholarly opinion against the neo-Malthusians. In his *Science* article, Simon not only singled out *The Population Bomb* as the "most influential" text in the neo-Malthusian oeuvre, but its title alluded to Ehrlich's co-authored book *Population, Resources, Environment* (1970). The gambit worked. "Resources, Population, Environment" not only provoked the largest batch of letters in *Science*'s history (including the cutting response from Ehrlich and his collaborators), but it also led to Ehrlich and Simon's famous bet. The two rivals could agree to terms in 1980 because they both believed the market would register environmental disruptions through higher prices, much like a seismograph translates subterranean tremors to ink and paper. The bet, however, revealed to Ehrlich that "prices of metals really don't have much to do with environmental quality."[43] While it catapulted Simon and neoliberal cornucopianism to prominence, Ehrlich's defeat represented a broader turn in the neo-Malthusians' fortunes.

A chastened but still defiant Ehrlich challenged Simon to a different kind of bet in 1995. He made the offer in response to Simon's boast in Ehrlich's hometown newspaper—and on Earth Day no less—that "every measure of material and environmental welfare in the United States and in the world has improved rather than deteriorated."[44] Ehrlich's terms were a ten-year wager on fifteen indicators based on physical rather than monetary attributes (e.g., atmospheric CO_2, average temperature, availability of firewood in poor countries, fisheries landings, and the hole in the ozone layer). Simon haggled with Ehrlich to structure the bet that resources services would improve (e.g., lower energy prices in general rather than plentiful firewood). Unlike the 1980 bet, neither side yielded. "He can take the bet or not," Simon told the *San Francisco Chronicle*, "and if he doesn't want to take the bet then he can shut up about people not being willing to bet him on important trends."[45] Ehrlich, who would have easily won this second bet, had learnt the lesson that the market was the wrong institution to comprehend and control humanity's interchange with nature. Instead of relying on prices, it was necessary to use many incommensurate physical metrics so that the economy could be consciously constrained within ecological limits.

Despite Ehrlich's defeat in the 1990 bet, the rapid decline of the neo-Malthusian movement is surprising, given that many of Ehrlich's predictions have proven correct. After all, millions *did* die from food shortages after 1968, with the average toll in recent years being 3.5 million young children perishing annually from malnutrition (the same demographic Simon dismissed in "Resources, Population, Environment").[46] Global wildlife biomass has declined by half in the last fifty years, and countless species have become extinct.[47] Simon may have won the bet, but this hardly proved that market pressure tends to make natural resources cheaper. Researchers of a recent study on the Ehrlich-Simon bet found that Simon would have lost most of the ten-year intervals in the twentieth and early twenty-first centuries, concluding that "it is better to be lucky than good."[48] It is a small irony that when neoliberal governments precipitated recessions to crush the labour movement, they inadvertently helped Simon win his bet.

The spectre of 1970s-style shortages reappeared in the 2000s, as petroleum peaked at $145 per barrel and the cost for a bushel of corn more than tripled. This energy crisis was linked to the transition from conventional to nonconventional petroleum. Veterans of the neo-Malthusian wave of the 1960s and their descendants heralded this crisis as the arrival of "peak oil."

That term was inspired by Marion King Hubbert, a geologist at Shell Oil Company, who made two bold predictions at the American Petroleum Institute conference in 1956. He aggregated the S-curves (i.e., rates of growth) of many historical oil fields in the US to get a sense of the trajectory of the national and global industry. First, he augured that US oil conventional production would peak in 1970—much earlier than any of his contemporaries expected—while global production would follow in 2000.[49] The repercussion of the first peak was that the US was no longer a "swing producer" able to make up for shortages elsewhere, as it had during the wars in the Middle East during 1956 and 1967. Hubbert's second peak arrived in the early 2000s, when global conventional petroleum production stagnated even as prices shot up.[50] Eventually, additional nonconventional production from deep-sea rigs, hydraulic fracturing (i.e., "fracking"), and the tar sands allowed global petroleum output to increase and moderate prices.

Simon, who died in 1998, did not live to see the nonconventional boom in the 2000s and 2010s, but his neoliberal comrades were convinced that it vindicated their fallen champion.[51] In 2005, journalist and arch-Simonian John Tierney made a $5,000 bet with Matthew Simmons (a "peak oil" defender and banker) whether the price of oil would triple by 2010. When the bet was called, Tierney had easily won and in the pages of the *New York Times* he attributed his victory less to the recent recession than to the "Cornucopian feast" of "giant new oil fields [. . .] off the coasts of Africa and Brazil [and] the new oil sands projects in Canada."[52] Similarly, in 2005 the Heartland Institute trumpeted the "Julian Simon effect," where higher prices were rapidly transforming Canada's tar sands into a viable industry. Simon's enduring legacy is affirmed every time a resourceship-based book is published, and there are many: Bjørn Lomborg's *The Sceptical Environmentalist* (2000 [1998]), Matt Ridley's *The Rational Optimist* (2010), Johan Norberg's *Progress* (2016), and Marian Tupy and Gale Pooley's *Superabundance* (2022).[53] The nonconventional boom seemed to secure the neoliberals a paradoxical victory over the neo-Malthusians, as the superabundance of fossil fuels has become a greater problem than their scarcity. This hints at the more fundamental problem: whether capitalism is more resilient than Earth.

Resourceship was never the neoliberals' most elegant intellectual construction, but its ramshackle mix of interwar German institutionalism, 1960s revisionist economic history, and neoliberal epistemology has attracted adherents across the political spectrum. Neoclassical economists now are far less worried about finding a "backstop" energy source than they were in the 1970s. Environmental computer simulations, including present-day integrated assessment models, that descend from *The Limits to Growth*'s projection in 1972 have strayed from their neo-Malthusian roots by incorporating "endogenous technological change" into their assumptions, thus perpetually postponing ecological doomsday.[54] Unlike the modellers of fifty years ago, today's environmental engineers have little interest in simulating a controlled, post-capitalist economy and instead endeavour to find ever more improbable ways of bending the biosphere with so-called negative emission technologies to fit capital's needs.[55] We have all become Simonians.

After the damp squib of the 2000s "peak oil" scare, even progressive movements of various stripes have come to embrace at least some tenets of resourceship. Amongst environmentalists, the "ecomodernists" Ted Nordhaus and Michael Shellenberger are Simon's most studious students as they share his belief that *more* rather than *less* market dynamism is necessary to achieve the technological breakthroughs needed to overcome the environmental crisis. The two come to prominence after penning "The Death of Environmentalism" (2004), an anodyne critique of the contemporary environmentalist strategy of fighting climate change, and set up the Breakthrough Institute (BI) in 2007. Soon, however, the BI became a hub of pro-growth economists, Promethean socialists, neoliberals, and geoengineers. Some of this motley crew joined Nordhaus and Shellenberger in writing their "Ecomodernist Manifesto" (2015), which openly abjured twentieth-century neo-Malthusianism and called for a "good Anthropocene" based on "decoupling" economic growth from material throughput.[56] Some science and technology studies scholars have joined the Simonian fold too. Bruno Latour published an essay on the BI website, where he exhorted readers to "love their monsters" and to never "stop innovating, inventing, creating, and intervening." Perhaps Simon's most influential and astute intellectual heir is William MacAskill, a leading philosopher of the Effective Altruism (EA) movement. He and other young utilitarians believed that through measuring "qalys" (quality-adjusted life years) and rigorous ranking, they could find the most effective ways of relieving suffering in the world. At first much of their work was directed toward buying anti-malaria nets and supporting animal rights, but eventually it turned towards "long-termist" issues like interplanetary travel and artificial intelligence. MacAskill's vision of the far future eerily mirrors Simon's neoliberal fantasy of a market-coordinated society spread amongst the stars. MacAskill and his acolytes have clearly grasped Simon's profound insight that capital must be able to escape the Earth before it outgrows its terrestrial confines. Given the frantic activity of billionaires like Elon Musk and Jeff Bezos in recent years to create a profitable space industry, these agents of capital must realize that time is running short.[57]

"Anyone who believes that exponential growth can go on forever in a finite world is either a madman or an economist," quipped the environmental economist Kenneth Boulding in 1973.[58] Yet, environmentalists do not seem to realize that Simon and other neoliberals *agree* that endless economic growth on Earth is impossible—this is why from the beginning resourceship was based on the assumption that capitalism *must* be an inter-galactic project. Spaceship Earth is too small a vehicle for an endlessly expanding entity like capital, which will eventually depend on uranium mined on "other planets."[59] While some Promethean Marxists and EA activists have accepted this, ecomodernists at the BI and neo-classical economists have proven more hesitant. The latter two groups believe that the solution to the environmental crisis is "de-coupling" pollution from economic growth, but Simon and other neoliberals have never deluded themselves in this regard. They prefer the multiplication of capital on many planets to the more fantastical goal of economic decoupling on a single world. Either capital dies along with its planetary host, or it escapes to the stars, or humanity successfully shackles the economy within planetary boundaries through planning based on non-monetary indicators. Increasingly, humanity as a whole has been betting on the unsubstantiated neoliberal faith in the market.

JULIAN L. SIMON
"Resources, Population, Environment: An Oversupply of False Bad News" (1980)*

In September 1977 *Newsweek* reported that "more than 100,000 West Africans perished of hunger" in the Sahel between 1968 and 1973 because of drought.† Upon inquiry, the writer of the account, Peter Gwynne, informed me that the estimate came from Kurt Waldheim's message to the United Nations' Desertification Conference. I therefore wrote to Waldheim asking for the source of the estimate.

Three mutually contradictory documents came back from the United Nations' Public Inquiries Unit: (i) Waldheim's message to the conference, saying, "Who can forget the horror of millions of men, women and children starving, with more than 100,000 dying, because of an ecological calamity that turned grazing land and farms into bleak desert?" (ii) A two-page excerpt from a memo by the U.N. Sahelian Office, dated 8 November 1974, saying, "It is not possible to calculate the present and future impact of this tragedy, on the populations. . . . Although precise figures are not available, indeed unobtainable . . . certainly there has been an extensive and tragic loss of life. . . ." (iii) A one-page memo written for the United Nations by Helen Ware, an Australian expert on African demography, who was a visiting fellow at the University of Ibadan in March 1975 when she wrote it. From calculations of the normal death rate for the area, together with "the highest death rate in any group of nomads" during the drought, she estimated "an absolute, and most improbable, upper limit [of] a hundred thousand. . . . Even as a maximum [this estimate] represents an unreal limit."

Ware's statement, which makes nonsense of Waldheim's well-publicized

---

*Julian L. Simon, "Resources, Population, Environment: An Oversupply of False Bad News," *Science* 208 (1980): 1431–37 (selected excerpts). Reprinted with permission from AAAS. The endnotes that follow on from these excerpts are from Simon's article.

†*Newsweek*, 19 September 1977, 80.

"Resources, Population, Environment" (1980)

assessment, was on page one of a document written for the United Nations well before the Desertification Conference. Apparently it was the only calculation the United Nations had, and it was grossly misinterpreted. [...]

A recent summary of the scientific evidence on the drought's effects by John Caldwell, a demographer who was familiar with the area prior to the drought and spent 1973 there, says, "One cannot certainly identify the existence of the drought in the vital statistics . . . nutritional levels, although poor, were similar to those found before the drought in other parts of Africa. The only possible exception was that of very young children."*

This is an example of a common phenomenon: Bad news about population growth, natural resources, and the environment that is based on flimsy evidence or no evidence at all is published widely in the face of contradictory evidence. [...]

SOME OTHER MYTHS ABOUT POPULATION AND RESOURCES

Here are some other examples of publicized, false, bad news and the unpublicized, good-news truth:

Statement: The food situation in less-developed countries is worsening. "Serious World Food Gap Is Seen Over the Long Run" is a typical *New York Times* headline.

Perhaps most influential in furthering that idea was Paul Ehrlich's best-selling book *The Population Bomb*, which begins: "The battle to feed all of humanity is over. In the 1970's the world will undergo famines—hundreds of millions of people are going to starve to death."† Many writers view the situation as so threatening that they call for strong measures to restrict population

---

*J. Caldwell, in *Drought in Africa* No. 2 (*African Environment Special Report No. 6*), D. Dalby, R. J. H. Church, F. Bezzaz, Eds. (UNEP-IDEP-SIDA, London, 1977), 93–100. For a full and judicious assessment of the situation and the area see Caldwell's *The Sahelian Drought and Its Demographic Implications* (Overseas Liaison Committee, American Council on Education, Washington, D.C., 1975).

†P. Ehrlich, *The Population Bomb* (Ballantine, New York, 1968), xi.

growth—"compulsion if voluntary methods fail," as Ehrlich put it.* Some, such as Paul and William Paddock, authors of the 1967 book *Famine—1975!*, find warrant in these assertions for such policies as "triage—letting the least fit die in order to save the more robust victims of hunger."† "My [one of the Paddocks] own opinion as the triage classification of these sample nations is: Haiti, Can't-be-saved; Egypt, Can't-be-saved; The Gambia, Walking Wounded; Tunisia, Should Receive Food; Libya, Walking Wounded; India, Can't-be-saved; Pakistan, Should Receive Food."‡

Fact: Per capita food production has been increasing at roughly I percent yearly—25 percent during the last quarter century. [. . .] Even in less-developed countries food production has increased substantially. World food stocks are high now, and even India has large amounts of food in storage. In the United States farmers are worrying about disaster from too much food.

Some countries have done far worse than the average, and have even had declining production, often because of war or political upheaval. And progress in food production has not been steady. But there has been no year, or series of years, so bad as to support a conclusion of long-term retrogression. Some readers might wonder whether my assertions are overly influenced by recent events, but the first draft of this material, for publication in my technical book,§ was written in 1971 and 1972, when food production was having its worst time in recent decades.

What about the data the other fellows quote to support their worried forecasts? In simple fact there are no other basic data. [. . .] Of course the data are less reliable than one would like; economic data usually are. But these are the only official data, and data that would show a worsening trend in recent decades simply do not exist. [. . .]

---

*Ibid.

†*Newsweek*, 11 November 1974, 16.

‡P. Paddock, *Famine—1975!* (Little, Brown: Boston, 1967), 222.

§Julian Simon, *The Economics of Population Growth* (Princeton: Princeton University Press, 1977).

"Resources, Population, Environment" (1980)

Statement: We are running out of natural resources and raw materials. "Entering an age of scarcity" is such a commonplace that it is simply assumed and asserted in public discussion by people ranging from B. F. Skinner to Solzhenitzyn.*

Response: The only meaningful measure of scarcity in peacetime is the cost of the good in question.† The cost trends of almost every natural resource—whether measured in labor time required to produce the energy, in production costs, in the proportion of our incomes spent for energy, or even in the price relative to other consumer goods—have been downward over the course of recorded history.

An hour's work in the United States has bought increasingly more of copper, wheat, and oil (representative and important raw materials) from 1800 to the present. And the same trend has almost surely held throughout human history. Calculations of expenditures for raw materials as a proportion of total family budgets make the same point even more strongly. These trends imply that the raw materials have been getting increasingly available and less scarce relative to the most important and most fundamental element of life, human work time. The prices of raw materials have even been falling relative to consumer goods and the Consumer Price Index. All the items in the Consumer Price Index have been produced with increasing ef-

---

*G. Homans, *Harv. Mag.* July-August 1977, p. 58; A. Solzhenitsyn, quoted in *Newsweek*, 18 March 1974, 122.

†H. J. Barnett and C. Morse [*Scarcity and Growth* (Johns Hopkins Press, Baltimore, 1963)] give the classic argument for this point of view, accompanied by a wealth of data. My discussion was inspired by their treatment and follows in their spirit, which in turn has roots in the Paley Commission of the early 1950s and in J. S. Davis, *J. Polit. Econ.* 61, 369 (1953). The data in *Scarcity and Growth* cover 1870 to 1957. Barnett has recently extended his analysis from 1957 to 1970 and found that the downward trends in real costs of extractive materials continue [H. J. Barnett, in *Scarcity and Growth Reconsidered*, V. K. Smith, ed. (Resources for the Future, Washington, D.C., in press]. A provocative but convincing technologically based argument for continuation of these downward cost trends for minerals is H. E. Goeller and Alvin M. Weinberg, *Science* 191, 683 (1976).

ficiency in terms of labor and capital over the years, but the decrease in cost of raw materials has been even greater than that of other goods, a very strong demonstration of progressively decreasing scarcity and increasing availability of raw materials.

The relative fall in the prices of raw materials understates the positive trend, because as consumers we are interested in the services we get from the raw materials rather than the raw materials themselves. And we have learned to use less of given raw materials for given purposes. as well as to substitute cheaper materials to get the same services. Consider a long-ago copper pot for cooking. The consumer is interested in a container which can be put over heat. After iron and aluminum were discovered, quite satisfactory cooking pots—almost as good as, or perhaps better than, pots of copper—could be made of those materials. The cost that interests us is the cost of providing the cooking service, rather than the cost of copper.

A dramatic example of how the service that copper renders can be supplied much more cheaply by a substitute process: A single communications satellite in space provides intercontinental telephone connections that would otherwise require thousands of tons of copper.

Statement: Energy is getting scarcer.

Response: The facts about the cost of energy are much the same as the facts about other raw materials. The new strength of the OPEC cartel to control oil price obscures the cost of production. But the production cost of a barrel of oil has not risen, and probably has fallen, in deflated dollars; even after the "oil crisis" of 1973 it was still $0.05 to $0.15 per barrel in the Persian Gulf, which was perhaps a hundredth of the market price.* It is reasonable to expect that eventually the price of oil will again return nearer its economic cost of production, and the long-run downward trend in the price of oil will resume its course.

The price of electricity is an interesting measure of the consumer cost of energy, and it is largely unaffected by cartels and politics (though the price of

---

*Jerusalem Post, 3 January 1978, p. 5; M. Zonis, *Univ. Chicago Mag.*, March 1976, 14.

"Resources, Population, Environment" (1980)

electricity did rise after 1973 because all energy sources, including coal and uranium, jumped in price when the price of oil went up, on account of the improved market power of coal and uranium suppliers). But the long-run cost of electricity clearly has been downward.

In short, the data show that energy has not been getting scarcer in basic economic terms, but rather has been getting more plentiful.

Statement: The supplies of natural resources are finite. This apparently self-evident proposition is the starting point and the all-determining assumption of such models as *The Limits to Growth* and of much popular discussion.

Response: Incredible as it may seem at first, the term "finite" is not only inappropriate but is downright misleading in the context of natural resources, from both the practical and the philosophical points of view. As with so many of the important arguments in this world, this one is "just semantic." Yet the semantics of resource scarcity muddle public discussion and bring about wrongheaded policy decisions.

A definition of resource quantity must be operational to be useful. It must tell us how the quantity of the resource that might be available in the future could be calculated. But the future quantities of a natural resource such as copper cannot be calculated even in principle, because of new lodes, new methods of mining copper, and variations in grades of copper lodes; because copper can be made from other metals; and because of the vagueness of the boundaries within which copper might be found—including the sea, and other planets. Even less possible is a reasonable calculation of the amount of future services of the sort we are now accustomed to get from copper, because of recycling and because of the substitution of other materials for copper, as in the case of the communications satellite.

Even the total weight of the earth is not a theoretical limit to the amount of copper that might be available to earthlings in the future. Only the total weight of the universe—if that term has a useful meaning here—would be such a theoretical limit, and I don't think anyone would like to argue the meaningfulness of "finite" in that context.

With respect to energy, it is particularly obvious that the earth does not bound the quantity available to us; our sun (and perhaps other suns) is our basic source of energy in the long run, from vegetation (including fossilized vegetation) as well as from solar energy. As to the practical finiteness and scarcity of resources—that brings us back to cost and price, and by these measures history shows progressively decreasing rather than increasing scarcity.

Why does the word "finite" catch us up? That is an interesting question in psychology, education, and philosophy; unfortunately, there is no space to explore it here.

In summary, because we find new lodes, invent better production methods, and discover new substitutes, the ultimate constraint upon our capacity to enjoy unlimited raw materials at acceptable prices is knowledge. And the source of knowledge is the human mind. Ultimately. then, the key constraint is human imagination and the exercise of educated skills. Hence an increase of human beings constitutes an addition to the crucial stock of resources, along with causing additional consumption of resources.

Statement: The old trends no longer apply. We are at a moment of discontinuity now.

Response: One cannot logically dispute assertions about present or impending discontinuity. And one can find mathematical techniques suggesting discontinuities that will be consistent with any trend data. We can say scientifically, however, that if in the past one had acted on the belief that the long-run price trend was upward rather than downward, one would have lost money on the average. [. . .]

WHY DO WE HEAR PHONY BAD NEWS?

Why do false statements of bad news dominate public discussion of these topics? Here are some speculations.

1. There is a funding incentive for scholars and institutions to produce bad news about population, resources, and the environment. The AID and the U.N.'s Fund for Population Activities disburse more than a hundred million dollars each year to bring about fertility decline. Much of this money goes to studies and publications that show why fertility decline is a good thing. There are no organizations that fund studies having the opposite aim.

"Resources, Population, Environment" (1980)

2. Bad news sells books, newspapers, and magazines; good news is not half so interesting. Is it a wonder that there are lots of bad-news best-sellers warning about pollution, population growth, and natural-resource depletion but none telling us the facts about improvement?

3. There are a host of possible psychological explanations for this phenomenon about which I am reluctant to speculate. But these two seem reasonably sure: (i) Many people have a propensity to compare the present and the future with an ideal state of affairs rather than with the past or with some other feasible state; the present and future inevitably look bad in such a comparison. (ii) The cumulative nature of exponential growth models has the power to seduce and bewitch.

4. Some publicize dire predictions in the idealistic belief that such warnings can mobilize institutions and individuals to make things even better; they think that nothing bad can come of such prophecies. But we should not shrug off false bad news as harmless exaggeration. There will be a loss of credibility for real threats as they arise, and loss of public trust in public communication. As Philip Handler. president of the National Academy of Sciences, testified to congressmen in the midst of the environmental panic of 1970: "The nations of the world may yet pay a dreadful price for the public behavior of scientists who depart from . . . fact to indulge . . . in hyperbole."*

The question, then, is: Who will tell us the good-and-true news? How will it be published for people to learn?

---

*P. Handler, interview in *U.S. News World Rep.*, 18 January 1971, 30.

CONCLUSION
Pluralistic Energy History in a Contested Epoch

DANIELA RUSS and **THOMAS TURNBULL**

ENERGY'S HISTORY **HAS SHOWCASED** the emergence of a more global and pluralistic school of energy history. We have done so in the hope of presenting a more representative picture of the energy historical past than the primarily Western-centered canon had previously done. By doing so, we learned that it is not enough to identify and translate material: the unevenness of globalization also manifests itself in the difficulty to ascertain copyright holders outside of conventional (mostly Western) publication channels. Within this system, republication is only a matter of money—beyond it, it can become a question of perseverance and luck. As acknowledged, it is the United States, the preeminent energy consumer in the period covered by this volume, that still dominates our analysis. Today, the United States has regained its title as the foremost fossil fuel producer in the world. Yet it is not without some new challengers. Qatar, Russia, South Africa, Australia, and China are themselves prodigious energy producers. However, alternative visions for energy-driven development abound, from full electrification to "green gas," degrowth, decoupling, and the contribution of alternative energies, from wind and solar to sugar ethanol. Changes are afoot. To gain insight into the dynamics and resistances that will configure this future, we have argued that both abstract and parochial forms of energy historical analysis must be abandoned in favor of an approach that seeks to understand both the locally situated motivations *and* the planetary consequences of past ideas and actions regarding energy consumption.

In providing sources from the preceding stage of energy history, our primary motivation was scholarly. We hope the more complex past, full of developmental and emancipatory visions and motivations, both realized and unrealized, of concepts, proven and unproven, gives those looking to achieve a different energy future a more nuanced picture of the past. The future of human energy use will of course be as complex and contradictory as that of the past. It is in this spirit of specificity that we hope students, researchers, and all who are interested in the historical agency of energy as both a physical force and as a concept will have found points of interest and encouragement to pursue energy-focused historical research. We also hope that the wider themes of this volume give the reader a more nuanced idea of the aspirations and fears that led to the formation of a largely fossil-fueled planet.

The genesis of this text lay in midst of the COVID-19 pandemic. Meeting online with contributors provided a welcome escape from an uncertain situation. Discussing texts and essays provided a forum to draw out overarching insights from a more pluralistic picture of the nascent discipline of energy history. As discussed in this volume's introduction, the Anthropocene, the contested idea that we have entered a new geological epoch, has been cast as a distinctly energy-historical idea. It emerged, so it is argued, from the geo-catalytic action of coal-powered steam engines, steamships, cotton mills, factories, and accelerated postwar economic growth in the West, the expansive energy infrastructures of the Communist sphere, what was once called the Second World, closely followed by fossil-fuelled developmentalism of the Global South, once known as the Third World. Rather than abstracting or reducing these events to mere quantitative or grand systemic "transitions," this volume gives insights into the hopes and fears that constructed the present.

Our first chapters addressed "Developmental Energies," texts that argued increased rates, control, and security of energy supply would provide twentieth-century nations with the productive and motive power to allow their continued self-determination in the wake of the First World War. Victor Seow's contribution described Japan's postwar anxieties about its seemingly inadequate coal resources and hopes that its Fuel Society would allow for increased efficiency of coal use. Fossil fuel use was seen as a means of national self-defense, rather than a means for colonial expansion. Shellen Wu provided a prehistory to China's present-day prodigious coal use, and akin to Seow, we see how Western-led surveys of coal seams both quelled developmental anxieties and raised fears the fuel might be ceded to avaricious Europeans. Hopes and fears coalesced into a domestic school of development-centered geology which fatefully "reduced coal to a numerical measure of progress." For countries without other affordances, it was alternative sources of energy that became enshrined with developmental potential. While Brazil had plenty of coal, as early as 1923, as Jennifer Eaglin showed, sugarcane ethanol became the hope of an ascendant Brazilian state. The idea was that the country's abundant land, cheap agricultural labor, and tropical climate could be put to work to propel state modernization and transportation. Finally, Antoine Missemer argued that in the post-war United States, the idea arose that economic growth could even be fueled by quotients of increased efficiency. What these chapters tell us is that the source of energy was largely irrelevant. The imperative was development amid geopolitical uncertainties, and the assumption was that this was best achieved with more or more efficient energy use. The result of this focus has been a range of externalities, from soot-caked cities to polluted waterways, by-products of the reductive pursuit of energy-driven growth. And yet, a salutary lesson emerges, as Missemer notes: the exact nature of the relationship between energy and growth remains uncertain. As such, it seems likely that other forms of prosperity that are less coupled to energy use are possible.

Energy has been seen not only as a means of state development but also as a means of emancipation from the imposition of colonial or hegemonic domination. In the "Emancipatory Energies" section, Elizabeth Chatterjee emphasized that colonialism was largely "technologically determined," and as such, plans to resist colonial subjugation, in the view of Indian physicist Meghnad Saha, would need countervailing energy-using technologies. Saha envisioned an alternative path to industrial and state independence, which would learn from the mistakes of coal-powered imperialists and make use of Indigenous sources of power, such as the flowing waters of the Damodar Valley. While Daniela Russ's chapter sets out another alternative vision, in which the Soviet Union would, through central planning, become distinguished by the efficiency of its use of locally available sources of power, from peat to coal, to water, and eventually to solar and nuclear power, all connected through a distributed but centralized electrical network. Infrastructure and ideology would harmonize. Largely unrealized and forgotten by the fall of the Soviet Union, the future envisioned by Krzhizhanovskii remains a potent vision of the possible advantages of a planned energy economy versus one shaped solely by competitive capitalism.

In Michael Dobson and Giuliano Garavini's contribution we revisited the oft-forgotten origins of perhaps the twentieth century's most effective anticolonial organization, the Organization of Petroleum Exporting Countries (OPEC) in the 1960s. The idea that oil-producing nations might use the West's appetite for oil to assert their developmental and nationalistic objectives, we learn, was the work of Venezuelan Juan Pablo Pérez Alfonzo. Their essay affirms how Pérez Alfonzo put forward a comprehensive system of "technically sophisticated, democratically grounded, anticolonial-internationalist, and conservationist political thought," which Western observers sought to denigrate and often racialize as a foreign "oil weapon." However, not all visions of energetic emancipation placed their faith in the state. In Lagos, as Damilola Adebayo argued, an African elite sought electrification akin to that illuminating Paris and London, but one vocal proponent, Fredrick William Dove, doubted the ability of the British colonial government to implement it and instead called on private industry to do so. In these contributions we see actors imbue energy technologies with a radical potential, a means to forge a path to independence. The path to a fossil-fueled planet was not driven solely by energetic colonialism but also by attempts at its resistance.

"Ideational Energies" is perhaps the most challenging set of essays in the volume, as it is focused on the historicity of the energy concept itself. In Laura Ann Twagira's contribution, she introduced readers to *nyama* via a Soudanese (today Malian) folk tale. Twagira argued that *nyama* is a vernacular energy-like concept that existed independently from Western European conceptions of energy but remained coherent with them. In pointing to the existence of energy-like concepts in non-Western contexts, we can credibly parochialize the Western European energy concept and ask what has been lost in the imposition of a singular concept on the many affordances of Earth's forces. Moreover, in Rebecca Wright's contribution we learned that the precepts of thermodynamics remained uncertain even in the heart of Western society. She returned us to philosopher William James and his public musings about mental energy, the ideational force that springs people into action. This unquantifiable and mysterious force, James argued, put paid to the idea that the quantity and rate of energy use was the primary determinant of social progress. Against simplistic theories of energetic determinism, James argued that the *organization* of energy use was more important. The application of mental and moral prudence in its expenditure was key. Thomas Turnbull's chapter followed through on this focus on the concept of energy. In the 1970s, US anthropologist Laura Nader observed the work of US engineers and physicists and concluded that, like any social group, they had developed cultural values that informed their conception of energy's societal role. In applying the principles of cultural relativism to US energy policy, her work opened a space to imagine alternative energy using societies. However, not all visions of a less determinate relationship between energy and society are oriented toward a more equitable future. In Troy Vettese's concluding chapter, we saw how, via a recontextualization of economist Julian Simon's concept of "resourceship," ongoing uncertainties about the indeterminacy of energy and resource limits remain a key aspect of an influential school of neoliberal economics that seeks to break free of earthly constraints. All these chapters point to one central fact: despite the supposedly lawlike nature of energy, its cultural reception is diffuse, uncertain, and open to contestation. Energy itself is subject to historical change.

What then do these many plans, arguments, and visions of more advantageous, emancipatory, or expansive uses of Earth's energies tell us if treated in totality? If the Anthropocene is taken seriously as both a geo-historical and a human historical period, whether or not it is formally recognized by the wider field of Earth scientists, the specific and localized intellectual histories of energy that have been presented here, and the increased rate and scale of human energy use that they contributed to or observed, have become folded into Earth's history. The utilization of the Sun, the trajectory of our climate, the diversity of our plant and animal species, and the composition of chemical and sedimentary cycles have been shaped by the energy histories that our protagonists described. Alternative developmental paths—roads not taken—could possibly have delayed things or perhaps altered Earth more subtly. But it appears that the direction of travel was, to some extent, rightly predicted by the energy-determinist Henry Adams. While the paths leading us to the present have been varied, on a planetary scale, entropy inescapably increases. The question, for those who compose the future, is how to advantageously, artfully, and more equitably make use of our planet's energetic disequilibria. The ongoing increases in entropy are not a condition that dictates the end of human history so much as one that describes the conditions within which an array of alternative futures may become possible.

Acknowledgments

We would like to thank all our contributors and readers for their sustained efforts in coming together (via video-conferencing software!), commenting on, and supportively reviewing each other's work, and for remaining enthusiastic during at times adverse circumstances.

For their generous institutional support, we would like to thank Jürgen Renn from the Max Planck Institute for the History of Science, as well as David Kaldewey and Doris Westhoff from the University of Bonn.

We also thank Lindy Divarci for her invaluable editorial support, and Thomas Behrendt for doggedly tracking down authorship rights.

Thanks also to Helge Wendt for inviting us to present the introduction at a colloquium.

This project received financial support from the Independent Social Research Foundation (ISRF) Small Grants Award, for which we are grateful.

Notes

Introduction

1. Béatrice Cointe and Héléne Guillemot, "A History of the 1.5°C Target," *WIREs Climate Change* 14, no. 3 (2023): 1–11.

2. Gurminder Bhambra and Peter Newell, "More Than a Metaphor: 'Climate Colonialism' in Perspective," *Global Social Challenges Journal* 2, no. 2 (2023): 179–87.

3. Adrian Lahoud, "Floating Bodies," in *Forensis: The Architecture of Public Truth*, eds. Eyal Weizman and Anselm Franke (Berlin: Sternberg Press, 2014), 508.

4. Vyoma Jha, "India and Climate Change: Old Traditions, New Strategies," *India Quarterly*, 78, no. 2 (2022): 280–96.

5. Vijaya Ramachandran, "Rich Countries' Climate Policies Are Colonialism in Green," *Foreign Policy*, November 3, 2021, https://foreignpolicy.com/2021/11/03/cop26-climate-colonialism-africa-norway-world-bank-oil-gas/; Hamza Hamouchene, *Dismantling Green Colonialism: Energy and Climate Justice in the Arab Region* (London: Pluto Press, 2023); for a study that analyzes the emergent inequalities without calling them "green colonialism," see Daniela Gabor und Ndongo Samba Sylla, "Derisking Developmentalism: A Tale of Green Hydrogen," *Development and Change* 54, no. 5 (2023): 1169–96.

6. Zhiwei Wang und Yongjun Huang, "Natural Resources and Trade-Adjusted Carbon Emissions in the BRICS: The Role of Clean Energy," *Resources Policy* 86 (2023): 104093.

7. Gregory Brew, *Petroleum and Progress in Iran: Oil, Development, and the Cold War* (Cambridge: Cambridge University Press, 2022); Elizabeth Chatterjee, "Reinventing State Capitalism in India: A View from the Energy Sector," *Contemporary South Asia* 25, no. 1 (2017): 85–100; Elizabeth Chatterjee, "The Asian Anthropocene: Electricity and Fossil Developmentalism," *The Journal of Asian Studies* 79, no. 1 (2020): 3–24; Antoine Acker, "A Different Story in the Anthropocene: Brazil's Post-Colonial Quest for Oil (1930–1975)," *Past & Present* 249, no. 1 (2020): 167–211.

8. Jeronim Perović, ed., *Cold War Energy: A Transnational History of Soviet Oil and Gas* (Cham: Palgrave Macmillan, 2017); Oscar Sanchez-Sibony, *The Soviet Union and the Construction of the Global Market: Energy and the Ascent of Finance in Cold War Europe, 1964–1971* (Cambridge: Cambridge University Press, 2023); Giuliano Garavini, "Completing Decolonization: The 1973 'Oil Shock' and the Struggle for Economic Rights," *The International History Review* 33, no. 3 (2011): 473–87.

9. Clapperton Chakanetsa Mavhunga and Helmuth Trischler, "Energy (and) Colonialism, Energy (In)Dependence: Africa, Europe, Greenland, North America," *RCC Perspectives*, no. 5 (2014), https://doi.org/10.5282/rcc/6554.

10. See, for instance, the format of Libby Robin, Sverker Sorlin, and Paul Warde, eds., *The Future of Nature: Documents of Global Change* (New Haven: Yale University Press, 2013); or more recently, in terms of their pursuit of cosmopolitanism, Alison Bashford, Emily Kern, and Adam Bobbette, eds., *New Earth Histories: Geo-Cosmologies and the Making of the Modern World* (Chicago: University of Chicago Press, 2023).

11. The classic reference is Thomas Kuhn "Energy Conservation as an Example of Simultaneous Discovery," in *Critical Problems in the History of Science: Proceedings of the Institute for the History of Science*, ed. Marshall Clagett (Madison: University of Wisconsin Press, 1959), 321–56; for a more contextual history, see Crosbie Smith, *The Science of Energy: A Cultural History of Energy Physics in Victorian Britain* (Chicago: University of Chicago Press, 1998); and on religion and energy, see Helge Kragh, *Entropic Creation: Religious Contexts of Thermodynamics and Cosmology* (Ashgate: Routledge, 2008).

12. Caleb Wellum, "The Use of Energy History," *Modern American History* 6, no. 2 (2023): 201–19.

13. Maxine Berg, *The Machinery Question and the Making of Political Economy, 1815–1848* (Cambridge: Cambridge University Press, 1982).

14. John Nef, *Rise of the British Coal Industry*, vol. 2 (London: George Routledge, 1932), 330.

15. For an overview (that is somewhat Western-centered), see Eugene A. Rosa und Gary E. Machlis, "Energetic Theories of Society: An Evaluative Review," *Sociological Inquiry* 53, nos. 2–3 (1983): 152–78.

16. Felix Auerbach, *Die Weltherrin und ihr Schatten: Ein Vortrag über Energie und Entropie* (Jena: Verlag von Gustav Fischer, 1913).

17. Henry Adams, *The Tendency of History* (New York: Book League of America, 1919), 5.

18. Keith Burich, "Henry Adams, the Second Law of Thermodynamics, and the Course of History," *Journal of the History of Ideas* 48, no. 3 (1987): 467–82.

19. Sebastian Felten and Renée Raphael, "Early Modern Resources: An Introduction," *Isis*, 114, no. 3 (2023): 599–603.

20. Turnbull, ch. 11, this book; and Nikolai Kardashev, "Transmission of Information by Extraterrestrial Civilizations," *Soviet Astronomy* 8, no. 2 (1964): 217–21.

21. Thomas Turnbull, "Energy, History, and the Humanities: Against a New Determinism," *History and Technology* 37, no. 2 (2021): 247–92.

22. Richard Hirsch and Christopher Jones, "History's Contributions to Energy Research and Policy," *Energy Research and Social Science* 1 (2014): 106–11; Paul Sabin, "'The Ultimate Environmental Dilemma': Making a Place for Historians in the Climate Change and Energy Debates," *Environmental History* 15, no. 1 (2010): 76–93.

23. Dipesh Chakrabarty, "The Climate of History: Four Theses," *Critical Inquiry* 35, no. 2 (2009): 197–222.

24. A zettajoule is 10^{21} joules. Jaia Syvitski et al., "Extraordinary Human Energy Consumption and Resultant Geological Impacts beginning around 1950 CE Initiated the Proposed Anthropocene Epoch," *Communications Earth and Environment* 1 (2021): 1–13.

25. William Jordy, *Henry Adams: Scientific Historian* (Hamden: Archon Books, 1970), 206.

26. Edward Anthony Wrigley, *Energy and the English Industrial Revolution* (Cambridge: Cambridge University Press, 2010); Andreas Malm, *Fossil Capital: The Rise of Steam Power and the Roots of Global Warming* (London: Verso, 2016).

27. Reinhart Koselleck, "Social History and Conceptual History," *International Journal of Politics, Culture, and Society* 2, 3 (1989): 308–25.

28. Rolf Peter Sieferle, *The Subterranean Forest: Energy Systems and the Industrial Revolution*, trans. Michael P. Osmann (Winwick: White Horse Press, 2010); Wrigley, *Energy and the English Industrial Revolution*.

29. Joel Mokyr, *The Lever of Riches: Technological Creativity and Economic Progress* (New York: Oxford University Press, 1992); Robert C. Allen, *The British Industrial Revolution in Global Perspective* (Cambridge: Cambridge University Press, 2009).

30. Malm, *Fossil Capital*.

31. While this volume focuses mainly on twentieth-century societies on the brink of or amid industrialization, energy history can look back to analyze organic energy systems. See Pekka Hämäläinen, "The Politics of Grass: European Expansion, Ecological Change, and Indigenous Power in the Southwest Borderlands," *The William and Mary Quarterly* 67, no. 2 (2010): 173–208; John Cropper, "The Sparrow Loves Millet, but Labors Not": Energy Use and Infrastructure in the Senegal Valley, 1450–1760," *History and Technology* 39, no. 1 (2023): 42–64, as starting points.

32. William Stanley Jevons, *The Coal Question: An Inquiry Concerning the Progress of the Nation and the Probable Exhaustion of Our Coal-Mines* (London: Macmillan, 1866), 146; Simon Pirani, *Burning Up: A Global History of Fossil Fuel Consumption* (London: Pluto Press, 2018), 12.

33. John Nef, *The Rise of the British Coal Industry*, vol. 1. (London: George Routledge, 1932), 243; Elmar Altvater, "The Social and Natural Environment of Fossil Capitalism," *Socialist Register* 42 (2007): 37–59; Malm, *Fossil Capital*, 321–22.

34. Michael Adas, *Machines as the Measure of Men: Science, Technology, and Ideologies of Western Dominance* (Ithaca: Cornell University Press, 1989), 134.

35. Among others, Sieferle, *The Subterranean Forest*.

36. Edward Wrigley, *Industrial Growth and Population Change: A Regional Study of the Coalfield Areas of North-West Europe in the Later Nineteenth Century* (Cambridge: Cambridge University Press, 1962), 31.

37. Jürgen Osterhammel, *The Transformation of the World: A Global History of the Nineteenth Century* (Princeton: Princeton University Press, 2014), 637–67.

38. Daniel Headrick, *The Tools of Empire: Technology and European Imperialism in the Nineteenth Century* (New York-Oxford: University of Oxford Press, 1981), 17–21; Adas, *Machines*, 203.

39. Bruce Podobnik, *Global Energy Shifts: Fostering Sustainability in a Turbulent*

Age (Philadelphia: Temple University Press, 2005), 29; Osterhammel, *Transformation of the World*, 62; Laleh Khalili, *Sinews of War and Trade: Shipping and Capitalism in the Arabian Peninsula* (London: Verso, 2020), 22.

40. For a selection, see Martin Lynn, "From Sail to Steam: The Impact of the Steamship Services on the British Palm Oil Trade with West Africa, 1850–1890," *Journal of African History* 30, no. 2 (1989): 227–45; Heidi Zogbaum, "The Steam Engine in Cuba's Sugar Industry, 1794–1860," *Journal of Iberian and Latin American Research* 8, no. 2 (2002): 37–60; Jennifer Tann, "Steam and Sugar: The Diffusion of the Stationary Steam Engine to the Caribbean Sugar Industry, 1770–1849," *History of Technology* 19 (1997): 63–84. Of course, Western energy technologies did not always replace indigenous ones; see Clive Dewey, *Steamboats on the Indus: The Limits of Western Technological Superiority in South Asia* (New Delhi: Oxford University Press, 2014).

41. Turnbull, "Energy, History, and the Humanities."

42. Robert Hartwell, "A Cycle of Economic Change in Imperial China: Coal and Iron in Northeast China, 750–1350," *Journal of the Economic and Social History of the Orient* 10, no. 1 (1967): 140.

43. Hartwell, "A Cycle." The value of Kaifeng trade at the end of the eleventh century was equivalent to well over 12,400,000 pounds (in 1702 values) whereas the imports and exports of London in 1711 were worth no more than 8,450,000 pounds (Hartwell, 144).

44. Hartwell, "A Cycle," 151–52.

45. Tim Wright, "An Economic Cycle in Imperial China? Revisiting Robert Hartwell on Iron and Coal," *Journal of Economic and Social History of the Orient* 50, no. 4 (2007): 398–423.

46. Kenneth Pomeranz, *The Great Divergence: China, Europe, and the Making of the Modern World Economy* (Princeton: Princeton University Press, 2000), 65; Prasannan Parthasarathi, *Why Europe Grew Rich and Asia Did Not: Global Economic Divergence, 1600–1850* (Cambridge: Cambridge University Press, 2011), 159–60.

47. Sujit Sivasundaram, "The Oils of Empire," in *Worlds of Natural History*, eds. Helen Curry, Nick Jardine, James Secord and Emma Spary (Cambridge: Cambridge University Press, 2018), 379–98; Daniel Yergin, *The Prize: The Epic Quest for Oil, Money and Power* (New York: Simon & Schuster, 1991).

48. Eric Williams, *Capitalism and Slavery* (London: Penguin, 2022).

49. Williams, *Capitalism and Slavery*, 81–84; on Williams' marginalisation, see Maxine Berg and Pat Hudson, *Slavery, Capitalism and the Industrial Revolution* (Cambridge: Polity, 2023), 37–41.

50. On Barak, *Powering Empire: How Coal Made the Middle East and Sparked Global Carbonization* (Oakland: University of California Press, 2020), 157; Crosbie Smith, *Coal, Steam and Ships: Engineering, Enterprise and Empire on the Nineteenth-Century Seas* (Cambridge: Cambridge University Press, 2018).

51. Steven Gray, *Steam Power and Sea Power: Coal, the Royal Navy, and the British Empire, c. 1870–1914* (London: Palgrave Macmillan, 2018).

52. Headrick, *Tools*, 154–55; Timothy Mitchell, *Carbon Democracy: Political Power in the Age of Oil* (London: Verso, 2011), 37.

53. Hubert Bonin, *History of the Suez Canal Company, 1858–1960: Between Controversy and Utility* (Geneva: Librairie Droz, 2010), 119; OEEC, *Europe's Need for Oil: Implications and Lessons for the Suez Crisis* (Paris: Organisation for European Economic Co-operation, 1958).

54. On Barak, "Outsourcing: Energy and Empire in the Age of Coal, 1820–1911," *International Journal of Middle Eastern Studies* 47, no. 3 (2015): 427.

55. Osterhammel, *Transformation of the World*, 713.

56. Peter Shulman, *Coal and Empire: The Birth of Energy Security in Industrial America* (Baltimore: Johns Hopkins University Press, 2015); Smith, *Coal, Steam and Ships*; Khalili, *Sinews of War and Trade*, 1.

57. Osterhammel, *Transformation of the World*, 62.

58. Podobnik, *Global Energy Shifts*, 14–15; Thomas Hughes, *Networks of Power: Electrification in Western Society, 1880–1930* (Baltimore: Johns Hopkins University Press, 1983); Jonathan Coopersmith, *The Electrification of Russia, 1880–1926* (Ithaca: Cornell University Press, 1992).

59. Paolo Malanima, "The Limiting Factor: Energy, Growth, and Divergence, 1820–1913," *The Economic History Review* 73, no. 2 (2020): 501–2.

60. Christophe Bonneuil and Jean-Baptiste Fressoz, *The Shock of the Anthropocene: The Earth, History and Us*, trans. David Fernbach (London: Verso, 2017), 241–42.

61. Daniela Russ, "'Socialism Is Not Just Built for a Hundred Years': Renewable Energy and Planetary Thought in the Early Soviet Union (1917–1945)," *Contemporary European History*, 31, no. 4 (2022): 491–508; Ariel Ron, "When Hay Was King: Energy History and Economic Nationalism in the Nineteenth-Century United States," *The American Historical Review* 128, no. 1 (2023): 117–213; Elizabeth Chatterjee, "The Poor Woman's Energy: Low-Modernist Solar Technologies and International Development, 1878–1966," *Journal of Global History* 13, no. 3 (2023): 439–60.

62. Jennifer Eaglin, *Sweet Fuel: A Political and Environmental History of Brazilian Ethanol* (Oxford: Oxford University Press, 2022), Appendices.

63. Victor Seow, *Carbon Technocracy: Energy Regimes in Modern East Asia* (Chicago: University of Chicago Press, 2021).

64. Shellen Wu, *Empires of Coal: China's Entry into the Modern World Order, 1860–1920* (Princeton: Princeton University Press, 2015).

65. Barak, "Outsourcing," 428; Hsien-Chun Wang, "Discovering Steam Power in China, 1840s–1860s," *Technology and Culture* 51, no. 1 (2010): 31–54.

66. Mitchell, *Carbon Democracy*, 158.

67. Giuliano Garavini, *The Rise and Fall of OPEC in the Twentieth Century* (Oxford: Oxford University Press. 2019), 8.

68. Elizabeth Chatterjee, "The Asian Anthropocene."

69. Cara New Daggett, *The Birth of Energy: Fossil Fuels, Thermodynamics, and the Politics of Work*, Elements (Durham: Duke University Press, 2019), 19; Daan Oostveen, "On the Concept of 'Energy' from a Transcultural Perspective," in *Energy Justice across Borders*, eds. Gunter Bombaerts et al. (Cham: Springer, 2020). Philosopher Yuk Hui points out that *qi* can also be translated as "gas," bringing the term back into the

ambit of energy; Yuk Hui, *The Question concerning Technology in China: An Essay in Cosmotechnics* (Falmouth: Urbanomic Media, 2016), 61.

70. Sean Hsiang-lin Lei, "Qi Transformation and the Steam Engine: The Incorporation of Western Anatomy and Re-conceptualisation of the Body in Nineteenth Century Chinese Medicine," *Asian Medicine* 7 (2012): 319–57.

71. On Barak, "Three Watersheds in the History of Energy," *Comparative Studies of South Asia, Africa and the Middle East* 34, no. 3 (2014): 440–53.

72. On power fantasies, see Dolores Greenberg, "Energy, Power, and Perceptions of Social Change in the Early Nineteenth Century," *The American Historical Review* 95, no. 3 (1990): 693–714. On twelfth-century Indian beliefs about perpetual motion, see Lynn White Jr., *Medieval Technology and Social Change* (Oxford: Oxford University Press, 1962), 130–34.

73. Germán Vergara, *Fueling Mexico: Energy and the Environment, 1850–1950* (Oxford: Oxford University Press, 2021); Diana Montaño, *Electrifying Mexico: Technology and the Transformation of a Modern City* (Austin: University of Texas Press, 2021).

74. Fredrik Meiton, *Electrical Palestine: Capital and Technology from Empire to Nation* (Oakland: University of California Press, 2019); Mark Driscoll, *The Whites Are the Enemies of Heaven: Climate Caucasianism and Asian Ecological Protection* (Durham: Duke University Press, 2020); Ying Ja Tan, *Recharging China in War and Revolution, 1882–1955* (Ithaca and London: Cornell University Press, 2021); Abena Dove Osseo-Asare, *Atomic Junction: Nuclear Power in Africa after Independence* (Cambridge: Cambridge University Press, 2019); Emily Brownell, "Reterritorializing the Future: Writing Environmental Histories of the Oil Crisis from Tanzania," *Environmental History* 27, no. 3 (2022): 747–71; Katayoun Shafiee, *Machineries of Oil: An Infrastructural History of BP in Iran* (Cambridge: MIT Press, 2023), *inter alia*.

75. Abigail Harrison Moore and Ruth Sandwell, *In a New Light: Histories of Women and Energy* (Montreal: McGill Queens University Press, 2021).

Chapter 1

1. "Meta: equilibrar o balanço," *O Estado de São Paulo* (hereafter *ESP*), November 15, 1975, 30; Tamás Szmrecsányi, *O planejamento da agroindústria canavieira do Brasil 1930/1975* (São Paulo: Editora Hucitec, 1979), 436–37.

2. For example, see Michael Barzelay, *The Politicized Market Economy: Alcohol in Brazil's Energy Strategy* (Berkeley: University of California Press, 1986).

3. The benefits of using oil rather than ethanol for combustion engines was arguably more a political than a technical question. Oil extraction was a very dangerous, labor- and land-intensive industry in the early twentieth century, and alternative fuels for automobiles were far more competitive than oil's present dominance would suggest. Bill Kovarik argues that petroleum interests were able to edge out the ethanol market in favor of oil by using biased scientific reports. Bill Kovarik, "Henry Ford, Charles Kettering and the Fuel of the Future," *Automotive History Review* 32 (1998): 7–27.

4. See Victor Seow, "Coal Will Be the Primary Fuel of the Future," ch. 2 of this

volume; and John Wirth, *The Politics of Brazilian Development, 1930–1954* (Stanford: Stanford University Press, 1970), 139; Joel Wolfe, *Autos and Progress: The Brazilian Search for Modernity* (Oxford: Oxford University Press, 2010), 28–33.

5. Regina Machado Leão, *Álcool: energia verde* (São Paulo: IQUAL, 2002), 89–90. Sabino de Oliveira published this research in the 1930s. Eduardo Sabino de Oliveira, *Álcool motor e motores a explosão* (Rio de Janeiro: Ministério do trabalho, indústria, e comercio, Instituto de tecnologia, 1937).

6. Moacyr Soares Pereira, *O problema do álcool-motor* (Rio de Janeiro: José Olympio Editora, 1942), 6–21; "Occorrencia do petroleo no estado de S. Catarina," *ESP*, October 15, 1955, 9. For comparison, see Hal Bernton et al., *The Forbidden Fuel: A History of Power Alcohol* (Lincoln: University of Nebraska Press, 1982).

7. Guiliano Garavini, *The Rise and Fall of OPEC in the Twentieth Century* (Oxford: Oxford University Press, 2019), 217–21.

8. Werner Baer, *The Brazilian Economy: Growth and Development*, 6th ed. (Boulder: Lynne Rienner, 2008), 77–80.

9. President Ernesto Geisel's statement at the Meeting of Ministers, September 10, 1974, Federative Republic of Brazil, *Second National Development Plan-II PND (1975–1979)* (Brasília, 1974), 3.

10. Moacyr Castro, "Pedro Biagi," in *Os desbravadores*, edited by Galeno Amorim (Ribeirão Preto: Palavra Mágica, 2001), 127–32; José Rubens da Silva, "De pai para filho," in *A Revista Santa Elisa: Uma história de trabalho e desenvolvimento* (Ribeirão Preto: MIC Editorial, 1996), 23–24.

11. Funds request letter from Maurilio Biagi Filho to the IAA president General Alvaro Tavares Carmo, June 17, 1974. BR RJANRIO IY A6.16 Caixa 0443; Szmrecsányi, *O planejamento*, 310. Maurilio Biagi Sr. helped author a notorious April 1974 report proposing a large-scale ethanol program to the National Petroleum Council, which produced extensive public debate.

12. Barzelay, *Politicized Market Economy*, 174.

13. Brazil and Iraq signed their first bilateral trade agreement in 1971. Their favorable relations allowed Iraq to sell Brazil petroleum at OPEC prices but without the "surcharges in force on the international market" to diminish some of the damage the higher prices would have on the Brazilian economy during the oil crisis. On Brazilian-Iraqi oil relations, see Seme Taleb Fares, "O Pragmatismo do Petróleo: as relações entre o Brasil e o Iraque," *Revista Brasileira de Políticas Internacionais* 50, no. 2 (2007): 127–45, quote on 137.

14. "CTA diz que pode utilizar o Programa Álcool-Motor no país," and "500 veículos andam apenas com álcool," *Jornal do Brasil*, May 29, 1978, 15.

15. For example, Lieutenant Coronel Sergio Ferolla, director of the Institute of Research and Development at the CTA. noted that Brazilian ownership of the alcohol motor patent provided the country "equal footing with international manufacturers." João Batista Olivi, "CTA ditará padrões para motor a álcool," *ESP*, April 27, 1979, 30.

16. Interview with Mário Garnero, December 11, 2013; "Um acordo com as fábricas para a produção de carros a álcool," *Gazeta Mercantil*, September 11, 1979.

17. Wolfe, *Autos and Progress*, 166; "País poderá até exportar carro popular," *ESP*, January 2, 1993, 21.

18. Even with the incentives provided for ethanol cars in the 1980s, few could afford this luxury, hence critics highlighted how few everyday Brazilians benefitted from the program. See Fernando Homem de Mello, *Proálcool, energia e transportes* (São Paulo: Enio Matheus Guazzelli, 1981), 160.

19. "A subida da montanha," *Veja* 708, March 31, 1982, 100–1; "Produtores de álcool acusam as montadores," *Folha de São Paulo*, June 10, 1983. See also, *O modelo energético brasileiro* (Brasília: Ministry of Mines and Energy, 1979), 40.

20. Nearly a third of the national fleet were ethanol-fueled cars, and ethanol continued to be mixed in the national fuel supply. Cley Scholz, "Bois invadem canaviais e poderá faltar álcool," *ESP*, November 11, 1986, 41; Ministry of Mines and Energy, *O modelo energético*, 41.

21. Anfavea, *Anuário da Indústria Automobilística Brasileira* (São Paulo: Anfavea, 2015), 59.

22. Shawn William Miller, *An Environmental History of Latin America* (New York: Cambridge University Press, 2007), 79–87.

23. Barzelay, *Politicized Market Economy*, 31.

24. Warren Dean, *With Broadax and Firebrand: The Destruction of the Brazilian Atlantic Forest* (Berkeley: University of California Press, 1995), 294.

25. Carlos Walter Porto-Gonçalves, "Implicações ecológicas e políticas do etanol," *Latin American Information Agency (ALAI)*, April 22, 2007, https://www.alainet.org/en/node/120689, accessed March 4, 2022. See also Herbert Klein and Francisco Vidal Luna, *Feeding the World: Brazil's Transformation into a Modern Agricultural Economy* (Cambridge: Cambridge University Press, 2018), ch. 3.

26. *Pro-álcool: mar de cana, mar de miséria* (Cuiabá: Associação de solidariedade às comunidades carentes de MT (Mato Grosso), Comissão Pastoral da Terra/regional de MT, and Centro de documentação terra e índio, 1984) 10–11.

27. "Aguas poluídas de nossos rios," *ESP*, December 22, 1953, 7; "Solicitada abertura de inquérito sobre a mortandade de peixes no rio Piracicaba," *ESP*, January 23, 1955, 15. See Municipal Law 393 of Piracicaba. Francisco Bergamin, "Usinas de cana e poluição dos rios," *ESP*, May 3, 1961, 42.

28. Governo do Estado de São Paulo, *Macrozoneamento das bacias dos rios Mogi Guaçu, Pardo, e Médio-Grande* (São Paulo: SMA/SAA/SEP, 199), 113; Eli Elias, "A incrível invasão dos pernilongos," *ESP*, September 29, 1979, 17; Josmar Verillo, "Usina emite comunicado oficial sobre vazamento de vinhaça," *Dourados News*, June 29, 2009, http://www.douradosnews.com.br/noticias/usina-emite-comunicado-oficial-sobre-vazamento-de-vinhaca-96e20babba89/362661/, accessed October 6, 2021.

29. According to reported data from Commissão Executivo Nacional de Álcool (CENAL), by the 1983–84 harvest year, Proàlcool created over 370,000 industrial and agricultural jobs. The vast majority would have been agricultural jobs. CENAL as published in Rubens Rodrigues dos Santos, "Proálcool alcançou os objetivos, afirma técnica," *ESP*, May 25, 1984, 24. See also Fernando Homem de Melo and Eduardo Giannetie da Fonseca, *Proálcool, energia, e transportes* (Sao Paulo: Pioneira, 1981), 82–94.

30. Peter Houtzager, "State and Unions in the Transformation of the Brazilian Countryside, 1964–1975," *Latin American Research Review* 33, no. 2 (1998): 107–13.

31. "O Proálcool, inflacionário," *ESP*, January 6, 1981, 24; Barzelay, *Politicized Market Economy*, 421–22.

32. "Produtores de álcool acusam as montadoras," *Folha de São Paulo*, June 10, 1983.

33. For example, Homem de Melo, *Proálcool, energia e transportes*, chapter 3; Fernando Homem de Melo, "Por que o álcool não é a melhor alternativa?," *Revista Exame* (São Paulo) 256 (1982): 102.

34. Interview with Élio Neves, May 26, 2020; see Jennifer Eaglin, *Sweet Fuel: A Political and Environmental History of Brazilian Ethanol* (Oxford: Oxford University Press, 2022), 135.

35. Only a year later, extreme working and living conditions drove those same workers to strike in the Ribeirão Preto region and forcefully demand benefits. After bringing production to a halt for three days, the strikes ultimately produced the first collective bargaining agreement between workers and employers in the country. See Eaglin, *Sweet Fuel*, chapter 5.

36. Jennifer Eaglin, "The Demise of the Brazilian Ethanol Program: Environmental and Economic Shocks, 1985–1990," *Environmental History* 24 (2019): 104–29.

37. Eduardo Massad, György M. Böhm, and Paulo H.N. Saldiva, "Ethanol Fuel Toxicity," in *Handbook of Hazardous Materials*, ed. Morton Corn (San Diego: Academic Press, 1993), 274; Pedro Jacobi, Denise Baena Segura, and Marianne Kjellén, "Governmental Responses to Air Pollution: Summary of a Study of the Implementation of *rodízio* in São Paulo," *Environment and Urbanization* 11, no. 1 (1999): 79–82.

38. Anfavea, *Anuário da Indústria Automobilística Brasileira*, 59

39. On labor and alternative energy technologies, see James Meek, "Who Holds the Welding Rod?," *London Review of Books* 43, no. 14 (2021): 1–19.

40. Clare Ribando Seekle and Brent Yacobucci, "Ethanol and Other Biofuels: Potential for U.S.-Brazil Energy Cooperation," CRS Report for Congress, September 7, 2007.

41. Jacques Moss, "What to Expect from Brazil's RenovaBio Programme," *KNect365 Energy*, September 5, 2018, https://knect365.com/energy/article/e5560843-78 a9-4034-81f7-25319afe103c/what-to-expect-from-brazils-renovabio-programme, accessed March 28, 2019.

42. On energy density as a barrier to transitioning from petroleum to biofuels, see Vaclav Smil, *Energy Transition: History, Requirements, Prospects* (Santa Barbara: Praeger, 2010), chapter 4.

43. José Coronado, "Bagaço aquece a indústria," *Química e derivados* 16, no. 190 (1982): 45, 50, 53; Galeno Amorim, "Simpósio discute alternativa para energia," *ESP*, August 18, 1989, 22.

44. Géraldine Kutas, "Bioplastics Further Unlock the Potential of Sugarcane," UNICA, November 21, 2011, accessed March 16, 2022.

45. On energy additions vs. transitions, see Christophe Bonneuil and Jean-Baptiste Fressoz, *The Shock of the Anthropocene* (London: Verso Books, 2017), ch. 5.

Chapter 2

1. This essay is based on portions of the third chapter of Victor Seow, *Carbon Technocracy: Energy Regimes in Modern East Asia* (Chicago: University of Chicago Press, 2022). I have provided references for instances in which something I mention in this essay is covered in greater detail in that text.

2. On Japanese resource anxieties and this framing of Japan as a "have-not country," see Eric Gordon Dinmore, "A Small Island Nation Poor in Resources: Natural and Human Resource Anxieties in Trans-World War II Japan" (Ph.D. dissertation, Princeton University, 2006); and Satō Jin, *'Motazaru kuni' no shigen ron: jinzoku kanō na kokudo o meguru mō hitotsu no chi* (Tokyo: University of Tokyo Press, 2011).

3. Anxieties over access to energy resources were one key driver of what I have termed "carbon technocracy," a modern regime of energy extraction predicated on "marshalling science and technology toward the exploitation of fossil fuels for statist ends" and further characterized by "an embrace of coal-fired development, a focus on heavy industrial expansion, a fixation on national autarky, an interest in labor-saving mechanization, a privileging of cheap energy, and a pegging of economic growth to increases in coal production and consumption." Seow, *Carbon Technocracy*, 4.

4. Seow, *Carbon Technocracy*, 41–42.

5. K. Okunaka, "Fuel Problem in Japan," in Y. Takenobu, ed., *The Japan Year Book, 1927* (Tokyo: Japan Year Book Office, 1927), supplement, 30.

6. On Japan's prewar petroleum industry, see Seow, *Carbon Technocracy*, 126–127.

7. David C. Evans and Mark R. Peattie, *Kaigun: Strategy, Tactics, and Technology in the Imperial Japanese Navy, 1887–1941* (Annapolis: Naval Institute Press, 1997), 184. A major factor driving the Japanese navy to convert its fleet to run entirely on oil was the desire to increase its operational capacities amid the restrictions placed upon its tonnage by the Washington Naval Treaty of 1922. See Seow, *Carbon Technocracy*, 125–126.

8. See, for example, "Waga sekiyu mondai to Bujun san yuko ketsugan no kachi," *Manshū nichinichi shinbun*, March 10, 1924.

9. J. Kurita, "Oil Consumption Index to Japan's Industry," *Trans-Pacific* 7, no. 1 (July 1922), 43–45.

10. Takenobu, *Japan Year Book, 1927*, 505; John R. Bradley and Donald W. Smith, *Fuel and Power in Japan* (Washington, D.C.: United States Government Printing Office, 1925), 7.

11. On the Japanese military's post–World War I embrace of self-sufficiency, see Michael A. Barnhart, *Japan Prepares for Total War: The Search for Economic Security, 1919–1941* (Ithaca: Cornell University Press, 1987), 22–49.

12. For more about the founding of the Fuel Society, see Seow, *Carbon Technocracy*, 132.

13. Quoted in Yoshimura Manji, "Nenryō kyōkai jūnen shi," *Nenryō kyōkai shi* 11, no. 10 (October 1932), 1196.

14. For more about the events the Fuel Society organized, including this fuel exhibition, see Seow, *Carbon Technocracy*, 133–34.

15. For more about this essay contest, see Seow, *Carbon Technocracy*, 134–135.

16. For more about the Fuel Society's efforts to get the Japanese government to enact a national fuel policy, see Seow, *Carbon Technocracy*, 135–36.

17. For more about Yoshimura, see Seow, *Carbon Technocracy*, 132–133.

18. On late-nineteenth- and early-twentieth-century Japanese discourses on "wealth and power," see Richard J. Samuels, *"Rich Nation, Strong Army": National Security and the Technological Transformation of Japan* (Ithaca: Cornell University Press, 1994), 1–32.

19. On the "population question" in interwar Japan, see Aya Homei, *Science for Governing Japan's Population* (Cambridge: Cambridge University Press, 2023), 92–97.

20. This part of Yoshimura's essay is not in the translated excerpt of this source book. For the original text, see Yoshimura Manji, "Waga kuni ni okeru nenryō mondai gaisetsu," *Nenryō kyōkai shi* 1, no. 1 (August 1922), 9.

21. For example, see Christopher F. Jones, *Routes of Power: Energy and Modern America* (Cambridge: Harvard University Press, 2014).

22. For Yoshimura's discussion of Jevons, see Seow, *Carbon Technocracy*, 130. On Jevons's *Coal Question*, see Fredrik Albritton Jonsson, "The Coal Question Before Jevons," *The Historical Journal* 63, no. 1 (February 2020): 107–126; and Victor Seow, *"The Coal Question* in the Age of Carbon," Joint Center for History and Economics, December 31, 2016, https://histecon.fas.harvard.edu/energyhistory/seow.html, accessed April 26, 2024.

23. "Imports," International Energy Agency, https://www.iea.org/reports/coal-information-overview/imports, accessed April 1, 2024; "Leading Oil Consuming Countries Worldwide in 2022," Statista, https://www.statista.com/statistics/271622/countries-with-the-highest-oil-consumption-in-2012/, accessed April 1, 2024.

24. Seow, *Carbon Technocracy*, 311.

Chapter 3

1. Rebecca Karl, *Staging the World: Chinese Nationalism at the Turn of the Twentieth Century* (Durham: Duke University Press, 2002).

2. Zhang Yiou (張軼歐) "Preface," *Dizhi huibao* (地質彙報, *Bulletin of the Geological Survey of China*), no. 1 (1919), unpaginated.

3. Jeffrey D. Sachs and Andrew M. Warner, "Natural Resources and Economic Development: The Curse of Natural Resources," *European Economic Review* 45, no. 4–6 (2001), 827–38.

4. Robert Hartwell, "A Cycle of Economic Change in Imperial China: Coal and Iron in Northeast China, 750–1350," *Journal of the Economic and Social History of the Orient* 10, no. 1 (1967): 102–59.

5. Kenneth, Pomeranz, *The Great Divergence: China, Europe, and the Making of the Modern World Economy* (Princeton: Princeton University Press, 2000).

6. Shellen Wu, *Empires of Coal: Fueling China's Entry into the Modern World Order, 1860–1920* (Stanford: Stanford University Press, 2015).

7. Magnus Fiskesjö, "Science across Borders, Johan Gunnar Andersson and Ding Wenjiang," in *Explorers and Scientists in China's Borderlands, 1880–1950*, eds. Denise

Glover, Steven Harell, and Charles F. McKhann (Seattle: University of Washington Press, 2011), 240–75.

8. *Dizhi huibao* 地質彙報 (*Bulletin of the Geological Survey of China*), no. 1 (1919), unpaginated.

9. Sociologists have examined the spread of science and the institutionalization of geological surveys globally. Quantitative examinations of this process provide startling visualisations of science's spread but do not examine who or how the transmission of science took place. Nevertheless, these studies provide a valuable basis for a historical analysis. See Evan Schofer, "The Global Institutionalization of Geological Science, 1800–1990," *American Sociological Review* 68, no. 5 (2003), 730–59.

10. For discussion of Richthofen's role in promoting Western interest in Chinese coal, see Shellen Wu, "The Search for Coal in the Age of Empires: Ferdinand von Richthofen's Odyssey in China, 1860–1920," *The American Historical Review* 119, no. 2 (2014), 339–62.

11. V. K. Ting [Ding Wenjiang], "Foreword," *Dizhi huibao* 地質彙報 (*Bulletin of the Geological Survey of China*), no. 1 (1919), unpaginated. Ding Wenjiang used the name V. K. Ting in his English-language writing.

12. Searching for Richthofen's name is further complicated by the multiple transliterations of his name in the nineteenth century.

13. Tim Wright, *Coal Mining in China's Economy and Society 1895–1937* (Cambridge: Cambridge University Press, 1984), 37.

14. Zheng Guanying, *Shengshi weiyan*, ed. Xie Lingmei (Zhengzhou: Zhongzhou guji chuban she, 1998), 381. The quote Zheng refers to comes from one of Ferdinand von Richthofen's letters to the Shanghai Chamber of Commerce. The original quote in Ferdinand von Richthofen, *Baron Richthofen's Letters, 1870-1872* (Shanghai: North China Herald Officer, 1903), 171.

15. For a more detailed discussion of these Republican-era geologists, see Grace Shen, *Unearthing the Nation: Modern Geology and Nationalism in Republican China* (Chicago: University of Chicago Press, 2014), 7–8.

16. See William Rowe, *Saving the World: Chen Hongmou and Elite Consciousness in Eighteenth-Century China* (Stanford: Stanford University Press, 2001), 2–3, and in more detail, 138–41.

17. Limin Bai, "*Gewu Zhizhi* and Curriculum Building," in *Re-Envisioning Chinese Education*, eds. Guoping Zhao and Zongyi Deng (Milton: Routledge, 2015), 55–73.

18. Wu, *Empires of Coal*, 50.

19. Grace Yen Shen, *Unearthing the Nation: Modern Geology and Nationalism in Republican China* (Chicago: University of Chicago Press, 2014).

20. Sigrid Schmalzer, *The People's Peking Man: Popular Science and Human Identity in Twentieth-Century China* (Chicago: University of Chicago Press, 2008).

21. V. K. Ting [Ding Wenjiang], "Foreword," *Dizhi huibao* (地質彙報, *Bulletin of the Geological Survey of China*), no. 1 (1919): 1, cited in Wu, *Empires of Coal*, 161–162.

22. Elizabeth Perry, *Anyuan: Mining China's Revolution Tradition* (Berkeley: University of California Press, 2012).

23. Andreas Malm, "China as Chimney of the World: The Fossil Capital Hypothesis," *Organization & Environment* 25, no. 2 (2012): 146–77.

24. Robert Finkelman and Linwei Tan, "The Health Impacts of Coal Use in China," *International Geology Review* 60, no. 5 (2019): 5–6.

25. Pengfei Xie, "Relocating Industry to Address Air Pollution in Beijing," The Nature of Cities Blog, https://www.thenatureofcities.com/2017/04/09/relocating-industry-address-air-pollution-beijing/, accessed August 9, 2023.

26. John A. Mathews and Hao Tan, *China's Renewable Energy Revolution* (London: Palgrave Macmillan, 2015).

27. Institute for Energy Research, "China and India Will Continue to Increase Oil and Gas Consumption, Paris Agreement Notwithstanding," May 26, 2017, http://instituteforenergyresearch.org/analysis/china-india-will-continue-increase-oil-coal-consumption-paris-agreement-notwithstanding/, accessed January 29, 2018.

28. IPCC, Working Group 1, *Climate Change 2021: The Physical Science Basis*, IPCC Sixth Assessment Report, https://www.ipcc.ch/report/ar6/wg1/, last accessed on August 31, 2021.

29. David Song-Pehamberger, "Green Tech Geopolitics: China and the Global Energy Transition," Foreign Brief: Geopolitical Risk Analysis, https://www.foreignbrief.com/analysis/green-tech-geopolitics/, accessed August 9, 2023.

30. Kristin Asdal provides an overview of theoretical concerns in environmental history in "The Problematic Nature of Nature: The Post-constructivist Challenge to Environmental History," *Environment and History* 42, no. 4 (2003): 60–74. The question of agency ties back to the work of anthropologist Anna Tsing and is also raised in Bruno Latour's essay "Objects too Have Agency," in *Reassembling the Social* (Oxford: Oxford University Press, 2005), 63–86.

31. Victor Seow, *Carbon Technocracy: Energy Regimes in Modern East Asia* (Chicago: University of Chicago Press, 2022), 232.

Chapter 4

1. Matthias Schmelzer, *The Hegemony of Growth. The OECD and the Making of the Economic Growth Paradigm* (Cambridge: Cambridge University Press, 2016).

2. Abbas Mardani et al., "Carbon Dioxide (CO_2) Emissions and Economic Growth: A Systematic Review of Two Decades of Research from 1995 to 2017," *Science of the Total Environment*, 649 (2019): 31–49; Dominik Wiedenhofer et al., "A Systematic Review of the Evidence on Decoupling of GDP, Resource Use and GHG Emissions, Part I: Bibliometric and Conceptual Mapping," *Environmental Research Letters* 15, no. 6 (2020): 033002.

3. Robert Fletcher and Crelis Rammelt, "Decoupling: A Key Fantasy of the Post-2015 Sustainable Development Agenda," *Globalizations*, 14, no. 3 (2017): 450–67.

4. See, for instance, Cutler J. Cleveland et al., "Energy and the U.S. Economy: A Biophysical Perspective," *Science* 225, no. 4665 (1984): 890–97; Balwinder S. Panesar and Richard C. Fluck, "Energy Productivity of a Production System: Analysis and Measurement," *Agricultural Systems* 43, no. 4 (1993): 415–37; Bernard C. Beaudreau,

Energy and Organization: Growth and Distribution Reexamined (Westport: Greenwood Press, 1998); Reiner Kümmel, Robert U. Ayres, and Dietmar Lindenberger, "Thermodynamic Laws, Economic Methods and the Productive Power of Energy," *Journal of Non-Equilibrium Thermodynamics* 35, no. 2 (2010): 145–79; Victor Court, "Énergie, EROI et Croissance Économique dans une Perspective de Long Terme," PhD thesis, Université Paris Nanterre, 2016.

5. Ernst R. Berndt, "Aggregate Energy, Efficiency, and Productivity Measurement," *Annual Review of Energy* 3 (1978): 225–73; Antoine Missemer and Franck Nadaud, "Energy as a Factor of Production: Historical Roots in the American Institutionalist Context," *Energy Economics* 86 (2020): 104706.

6. Samuel P. Hays, *Conservation and the Gospel of Efficiency* (Pittsburgh: University of Pittsburgh Press, 1959); and more recently, Ian Tyrell *Crisis of the Wasteful Nation: Empire and Conservation in Theodore Roosevelt's America* (Chicago: University of Chicago Press, 2015).

7. Gerald Alonzo Smith, "Natural Resource Economic Theory of the First Conservation Movement (1895–1927)," *History of Political Economy* 14, no. 4 (1982): 483–95; Jose Luis Ramos Gorostiza, "Ethics and Economics: From the Conservation Problem to the Sustainability Debate," *History of Economic Ideas* 11, no. 2 (2003): 31–52; Antoine Missemer, *Les Économistes et la Fin des Énergies Fossiles (1865–1931)* (Paris: Classiques Garnier, 2017).

8. W. G. Jensen, "The Importance of Energy in the First and Second World Wars," *The Historical Journal* 11, no. 3 (1968): 538–54; Malcom Rutherford, *The Institutionalist Movement in American Economics, 1918–1947* (Cambridge: Cambridge University Press, 2011), 8.

9. Joseph E. Pogue, *The Economics of Petroleum* (New York: Wiley & Sons, 1921); George Ward Stocking, *The Oil Industry and the Competitive System* (Clifton: Augustus M. Kelley, 1925).

10. John Ise, *The United States Forest Policy* (New Haven: Yale University Press, 1920); John Ise, *The United States Oil Policy* (New Haven: Yale University Press, 1926).

11. Marion Gaspard, Antoine Missemer and Thomas M. Mueller, "A Journey into Harold Hotelling's Economics," *Journal of Economic Literature* 62, no. 3 (2024): 1–37.

12. Rebecca Wright, Hiroki Shin, and Frank Trentmann, *From World Power Conference to World Energy Council. 90 Years of Energy Cooperation, 1923–2013* (London: World Energy Council, 2013); Daniela Russ, "Speaking for the 'World Power Economy': Electricity, Energo-Materialist Economics, and the World Energy Council (1924–78)," *Journal of Global History* 15, no. 2 (2020): 311–29.

13. Ernst R. Berndt, "From Technocracy to Net Energy Analysis: Engineers, Economists and Recurring Energy Theories of Value," Studies in Energy and the American Economy, Discussion Paper no. 11 (MIT, 1982); Daniela Russ and Thomas Turnbull, "Competing Powers: Engineers, Energetic Productivism, and the End of Empires," in *Competition in World Politics: Knowledge, Strategies, and Institutions*, ed. Daniela Russ and James Stafford (Bielefeld, Transcript: 2021), 183–210.

14. With the possible exception of Chester G. Gilbert and Joseph E. Pogue, *Amer-*

ica's Power Resources: The Economic Significance of Coal, Oil, and Water-Power (New York: Century, 1921).

15. Tryon participated, as a US delegate, in the 1924 World Power Conference (London) and in the 1929 World Engineering Congress (Tokyo).

16. Frederick G. Tryon, "An Index of Consumption of Fuels and Water Power," *Journal of the American Statistical Association* 22, no. 159 (1927): 271.

17. On the history of the US Geological Survey regarding energy resources and American supremacy, see Peter A. Shulman, *Coal & Empire. The Birth of Energy Security in Industrial America* (Baltimore: Johns Hopkins University Press, 2015); Roger J. Stern, "Oil Scarcity Ideology in US Foreign Policy, 1908–97," *Security Studies* 25, no. 2 (2016): 214–57. Tryon did not, however, specifically address this international dimension of energy resources.

18. On the history of the Brookings Institution, see Donald T. Critchlow, *The Brookings Institution, 1916–1952: Expertise and the Public Interest in a Democratic Society* (DeKalb: Northern Illinois University Press, 1985).

19. Harold G. Moulton, *The Brookings Institution, Report of the President for the Year Ending June Thirtieth 1929* (Washington, D.C.: Brookings Institution, 1929), 19.

20. Isador Lubin, *Miners' Wages and the Cost of Coal: An Inquiry into the Wages System in the Bituminous Coal Industry and Its Effects on Coal Costs and Coal Conservation* (New York: McGraw-Hill, 1924); Walton Hale Hamilton and Helen Russell Wright, *The Case of Bituminous Coal* (New York: Macmillan, 1925); Isador Lubin and Helen Everett, *The British Coal Dilemma* (New York: Macmillan, 1927); Harold G. Moulton, Charles S. Morgan, and Adah L. Lee, *The St. Lawrence Navigation and Power Project* (Washington, D.C.: Brookings Institution, 1929).

21. Letter from Tryon to Moulton, Dec. 4, 1926, Frederick G. Tryon papers, administrative record, Brookings Institution Library.

22. For instance, William H. Young, *Sources of Coal and Types of Stokers and Burners Used by Electric Public Utility Power Plants* (Washington, D.C.: Brookings Institution, 1930); Frederick G. Tryon and H. O. Rogers, "Statistical Studies of Progress in Fuel Efficiency," in *Transactions Second World Power Conference (Berlin)*, ed. F. zur Nedden and C. T. Kromer (Berlin: VDI-Verlag, 1930), 343–65; Frederick G. Tryon and Edwin C. Eckel, eds., *Mineral Economics. Lectures under the Auspices of the Brookings Institution* (New York and London: McGraw-Hill, 1932); Edwin G. Nourse et al., *America's Capacity to Produce* (Washington, D.C.: Brookings Institution, 1934).

23. Tryon, "An Index of Consumption," 271.

24. Gilbert and Pogue, *America's Power Resources*; see Frederick G. Tryon, "Review of 'The Economic Aspects of Geology' by C. K. Leith and 'America's Power Resources: The Economic Significance of Coal, Oil, and Water-Power' by C. G. Gilbert and J. E. Pogue," *Journal of the American Statistical Association* 18, no. 137 (1922): 144–46.

25. Carroll Roop Daugherty, "The Development of Horsepower Equipment in the United States," in *Power Capacity and Production in the United States*, ed. Carroll Roop Daugherty, Albert Howard Horton, and Royal William Davenport (Washington, D.C.: Government Printing Office, 1928), 5–112; Carroll Roop Daugherty, "Horse-

power Equipment in the United States, 1869–1929," *American Economic Review* 23, no. 3 (1933): 428–40.

26. Harry Smith, "The Origin of the Horsepower Unit," *American Journal of Physics* 4, no. 3 (1936): 120–22.

27. Rutherford, *Institutionalist Movement*; on institutionalism and energy see Thomas Turnbull, "From State to Market: A Transition in the Economics of Energy Resource Conservation," in *New Energies A History of Energy Transitions in Europe and North America*, ed. Stephen Gross and Andrew Needham (Pittsburgh: University of Pittsburgh Press, 2023), 135–38.

28. Tryon, "An Index of Consumption," 278, 280.

29. Tryon, "An Index of Consumption," 282. For his series on energy consumption, Tryon used sources common to Daugherty's. Regarding economic output, he relied on indexes of production from Edmund E. Day, "An Index of the Physical Volume of Production: IV. Agriculture, Mining, and Manufacturing Combines, 1899–1919," *Review of Economics and Statistics* 3, no. 1 (1921): 19–22; Carl Snyder, "An Index Number of Production: Discussion," *American Economic Review* 11, no. 1 (1921): 70–81; Carl Snyder, *Business Cycles and Business Measurements: Studies in Quantitative Economics* (New York: Macmillan, 1927); Walter W. Stewart, "An Index Number of Production," *American Economic Review* 11, no. 1 (1921): 57–70. To learn more about Tryon's sources and computation, see also Missemer and Nadaud, "Energy as a Factor of Production."

30. Tryon, "An Index of Consumption," 277–78.

31. Tryon, "An Index of Consumption," 278.

32. Berndt, "Aggregate Energy,"; Missemer and Nadaud, "Energy as a Factor of Production."

33. Tryon, "An Index of Consumption," 271.

34. Robert U. Ayres, "The Minimum Complexity of Endogenous Growth Models: The Role of Physical Resource Flows," *Energy* 26, no. 9 (2001): 817–38; Angeliki N. Menegaki, ed., *The Economics and Econometrics of the Energy-Growth Nexus* (London: Academic Press, 2018).

35. Tryon, "An Index of Consumption," 279. Tryon also had some methodological difficulties in explaining the lags between his series of data. This was mostly due to the lack of accurate statistical methods to address this issue at the time. Statistical treatments of distributed lags would appear only in the 1950s, with, for instance, Leendert M. Koyck, *Distributed Lags and Investment Analysis* (Amsterdam: North-Holland Publishing Company, 1954); Mark Nerlove, *Distributed Lags and Demand Analysis for Agriculture and Other Commodities* (Washington, D.C.: U.S. Department of Agriculture, 1958); Phoebus Dhrymes, *Distributed Lags. Problems of Estimation and Formulation* (Edinburgh: Oliver & Boyd, 1971).

36. Tryon, "An Index of Consumption," 273.

37. Émile Levasseur, *La Population Française : Histoire de la Population Française avant 1789 et Démographie de la France Comparée à celle des autres Nations au XIXè Siècle, Tome 3* (Paris: Arthur Rousseau, 1892), 74 (author's translation).

38. Bob Johnson, "Energy Slaves: Carbon Technologies, Climate Change, and the

Stratified History of the Fossil Economy," *American Quarterly* 68, no. 4 (2016): 955–79. See also Chatterjee, ch. 5 in this volume.

39. Gilbert and Pogue mention it, citing a calculation that energy sources provided "the labor of three billion hard-working slaves." See Gilbert and Pogue, *America's Power Resources*, 17.

40. Frederick G. Tryon, "The Effect of Competitive Conditions on Labor Relations in Coal Mining," *Annals of the American Academy of Political and Social Science* 111, no. 1 (1924): 82–95.

41. Frederick G. Tryon, "The Chances in Mining," in *Migration and Economic Opportunity. The Report of the Study of Population Redistribution*, ed. Carter Goodrich et al. (Philadelphia: University of Pennsylvania Press, 1936), 437.

42. Frederick G. Tryon et al., "The Mineral Industries," in *Technological Trends and National Policy Including the Social Implications of New Inventions: Report of the Subcommittee on Technology to the National Resources Committee*, ed. William F. Ogburn (Washington, D.C.: Government Printing Office, 1937), 174–76.

Chapter 5

1. See Edward Anthony Wrigley, *Continuity, Chance and Change: The Character of the Industrial Revolution in England* (Cambridge: Cambridge University Press, 1988); John R. McNeill, *Something New under the Sun: An Environmental History of the Twentieth-Century World* (London: Allen Lane, 2000); Rolf Peter Sieferle, *The Subterranean Forest: Energy Systems and the Industrial Revolution*, trans. Michael P. Osman (Cambridge: White Horse Press, 2001).

2. Ann Norton Greene, *Horses at Work: Harnessing Power in Industrial America* (Cambridge, MA: Harvard University Press, 2008). See also Ian Jared Miller and Paul Warde, "Energy Transitions as Environmental Events," *Environmental History* 24, no. 3 (2019): 463–533. In recognition of such continuities and contestations, some argue that we must reject linear, one-way terms like *electrification* as overly deterministic; Graeme Gooday, *Domesticating Electricity: Technology, Uncertainty and Gender, 1880–1914* (London: Pickering & Chatto, 2008), 1–2, 9–19.

3. For a critique of the "straitjacket histories of divergence or the Anthropocene," see Sujit Sivasundaram, "The Oils of Empire," in *Worlds of Natural History*, ed. Emma C. Spary et al. (Cambridge: Cambridge University Press, 2018), quote on 396.

4. Saha's works were collected in four volumes, the second of which was devoted to his copious writings on electricity, fuel, river management, industrialization, and planning; Santimay Chatterjee, ed., *Collected Works of Meghnad Saha* (Calcutta; Bombay: Saha Institute of Nuclear Physics; Orient Longman, 1982–1993), hereafter *CWMS*.

5. Pratik Chakrabarti, *Western Science in Modern India: Metropolitan Methods, Colonial Practices* (Delhi: Permanent Black, 2004), 287.

6. On this term, see also Missemer, ch. 4 in this volume.

7. This story is from Robert S. Anderson, *Nucleus and Nation: Scientists, International Networks, and Power in India* (Chicago: University of Chicago Press, 2010), 26.

8. Anderson, *Nucleus and Nation*, 24, 39. For a damning microhistory of the discrimination Saha faced, see Abha Sur, *Dispersed Radiance: Caste, Gender, and Modern Science in India* (New Delhi: Navayana, 2011). Another recent group biography argues that Saha was able "to overcome the caste barrier through education"; Somaditya Banerjee, *The Making of Modern Physics in Colonial India* (Abingdon: Routledge, 2020), 165. This complacent view is belied by a substantial body of scholarship that documents the persistence of caste discrimination in the highest echelons of Indian science and learning today, even as upper castes rush to declare the country post-caste; see most recently Ajantha Subramanian, *The Caste of Merit: Engineering Education in India* (Cambridge: Harvard University Press, 2019); Anna Ruddock, *Special Treatment: Student Doctors at the All India Institute of Medical Sciences* (Stanford: Stanford University Press, 2021).

9. David H. DeVorkin, "Quantum Physics and the Stars (IV): Meghnad Saha's Fate," *Journal for the History of Astronomy* 25, no. 3 (1994): 157.

10. Quoted in Anderson, *Nation and Nucleus*, 42.

11. In Berlin he worked in the laboratory of the chemist Walther Nernst, one of five future Nobel laureates who had spent the First World War working on poison gas for Germany. Rejecting London-based students in the war's aftermath, Nernst made an exception for Saha "because the last blow to the British Empire would come from India"; quote in Anderson, *Nation and Nucleus*, 35.

12. Saha, "Rethinking Our Future" (pamphlet, 1953), *CMWS* 2, 539–40.

13. Quoted in P. C. Mahalanobis in his own anniversary address, "Science and National Planning," *Sankhyā: The Indian Journal of Statistics* 20, no. 1/2 (1958): 69.

14. For example, National Planning Committee, *Power and Fuel: Report of the Sub-Committee*, ed. K. T. Shah (Bombay: Vora & Co., 1949), 24–25; Minoo Masani, *Our India* (Calcutta: Oxford University Press, 1940), 134–35.

15. Russian energy experts developed similar concepts, calculating the per capita "energy-equipment" (*energovooruzhenost'*) and "motor power outfit" of living labor. I am grateful to Daniela Russ for this observation; see her commentary in ch. 6 in this volume on Gleb Krzhizhanovskii, chief architect of Russia's famous GOELRO plan for state-led electrification. Saha was an admirer of Soviet electrification, and during his first visit to the country in 1945 he chatted with Krzhizhanovskii (in German) and inspected his institute's experiments on solar and wind energy; see *CWMS* vol. 4, 443–48. This is a useful reminder that modern energy thinking was always transnational.

16. *Science and Culture* 1 no. 1 (June 1935): 1. Cf. Sunil Purushotham, "World History in the Atomic Age: Past, Present and Future in the Political Thought of Jawaharlal Nehru," *Modern Intellectual History* 14, no. 3 (2017): 837–67.

17. Saha, "The Philosophy of Industrialization" (1938), *CWMS* vol. 4, 295–96.

18. On the imperial and increasingly racialized lineages of such ideas of technological superiority, see Michael Adas, *Machines as the Measure of Men: Science, Technology, and Ideologies of Western Dominance* (Ithaca: Cornell University Press, 1989). On evolutionary thought among earlier (but anti-state) Indian nationalists, see Shruti

Kapila, "Self, Spencer and *Swaraj*: Nationalist Thought and Critiques of Liberalism, 1890–1920," *Modern Intellectual History* 4, no. 1 (2007): 109–27.

19. See Benjamin Disraeli's novel *Coningsby*, for example, on the steam engines of Victorian Manchester: "A machine is a slave that neither brings nor bears degradation"; *Coningsby; Or, the New Generation* (London: Henry Colburn, 1844), vol. 2, 7. On the history of the body as machine, see Anson Rabinbach, *The Human Motor: Energy, Fatigue, and the Origins of Modernity* (Berkeley: University of California Press, 1992). This theme continued in the pages of *Science and Culture*, where the appropriate Indian diet was debated at length.

20. Bob Johnson, "Energy Slaves: Carbon Technologies, Climate Change, and the Stratified History of the Fossil Economy," *American Quarterly* 68, no. 4 (2016): 955–79.

21. Abha Sur, "Scientism and Social Justice: Meghnad Saha's Critique of the State of Science in India," *Historical Studies in the Physical & Biological Sciences* 33, no. 1 (2002): 98.

22. Sur, "Scientism and Social Justice," 96. This was one of the reasons for the Stakhanovite movement's successes in the Soviet Union, Saha argued: workers, not just the few supervisors, were encouraged to innovate; Saha, "Post-War Educational Development in India" (1944), *CWMS* vol. 3, 487.

23. Saha, "Technical Education," *Science and Culture* 13, no. 1 (1947): 43.

24. Johnson, "Energy Slaves," 956. On thermodynamics as colonial-capitalist ideology, see Cara New Daggett, *The Birth of Energy: Fossil Fuels, Thermodynamics, and the Politics of Work* (Durham: Duke University Press, 2019).

25. Jean-François Mouhot, "Past Connections and Present Similarities in Slave Ownership and Fossil Fuel Usage," *Climatic Change* 105 (2011): 329–55.

26. Aaron Bastani, *Fully Automated Luxury Communism: A Manifesto* (London: Verso, 2019).

27. Nikhil Menon, "Gandhi's Spinning Wheel: The Charkha and Its Regenerative Effects," *Journal of the History of Ideas* 81, no. 4 (2020): 643–62.

28. Richard B. Gregg, *Economics of Khaddar* (Triplicane, Madras: S. Ganesan, 1928), 19, 27. On the fascinating and quixotic Gregg, later a major influence on Martin Luther King Jr., see Joseph Kip Kosek, "Richard Gregg, Mohandas Gandhi, and the Strategy of Nonviolence," *Journal of American History* 91, no. 4 (2005): 1318–48.

29. Saha, Foreword to "If the War Comes" (1939), *CWMS* vol. 3, 385.

30. Saha, "Scientific Research in National Planning," *Science and Culture* 5, no. 11 (1940): 641.

31. On Britain's sluggish and haphazard electrification, see Leslie Hannah, *Electricity Before Nationalisation: A Study of the Development of the Electricity Supply Industry in Britain to 1948* (London: Macmillan, 1979).

32. On provincial variation in electrical investments under colonial rule, see Sunila Kale, "Structures of Power: Electrification in Colonial India," *Comparative Studies of South Asia, Africa and the Middle East* 34, no. 3: 454–75.

33. Saha, "Electricity—Its Use for the Public and for Industries" (1935), *CWMS* vol. 2, 185, 189.

250   Notes to Chapter 5

34. On how federalization shaped the slow and uneven pace of electrification, especially the uneconomic and ecologically ruinous subsidies secured by wealthy farmers in northwestern and southern states, see Sunila Kale, *Electrifying India: Regional Political Economies of Development* (Stanford: Stanford University Press, 2014); ch. 2 contains a useful summary of the Constituent Assembly debates. On the ecological consequences of this flow of cheap power to pump water for irrigation, see Sunil Amrith, *Unruly Waters: How Mountain Rivers and Monsoons Have Shaped South Asia's History* (London: Penguin, 2018), 256–88.

35. Kuntala Lahiri-Dutt, ed., *The Coal Nation: Histories, Ecologies and Politics of Coal in India* (Farnham: Ashgate, 2014).

36. Santimay Chatterjee, foreword, *CWMS* vol. 2, v.

37. Benjamin Siegel, *Hungry Nation: Food, Famine, and the Making of Modern India* (Cambridge: Cambridge University Press, 2018), 21–49.

38. Saha to Nehru, October 7, 1938, quoted in Sur, "Scientism and Social Justice," 93.

39. Carl Schmitt, *Roman Catholicism and Political Form*, trans. G. L. Ulmen (Westport: Greenwood Press, 1996), 13. On the strange parallels between the American and Soviet modes of energy exploitation and ecological harm, see Kate Brown, *Plutopia: Nuclear Families, Atomic Cities, and the Great Soviet and American Plutonium Disasters* (Oxford: Oxford University Press, 2013); Bathsheba Demuth, "The Walrus and the Bureaucrat: Energy, Ecology, and Making the State in the Russian and American Arctic, 1870–1950," *American Historical Review* 124, no. 2 (2019): 483–510.

40. David C. Engerman, *The Price of Aid: The Economic Cold War in India* (Cambridge: Harvard University Press, 2018).

41. Daniel Klingensmith, *"One Valley and a Thousand": Dams, Nationalism, and Development* (New Delhi: Oxford University Press, 2007).

42. Anil Agarwal and Sunita Narain, *Global Warming in an Unequal World: A Case of Environmental Colonialism* (New Delhi: Centre for Science and Environment, 1991), 1.

43. Agarwal and Narain, *Global Warming*, 11, 4, 9.

44. Arvind Subramanian, "India Is Right to Resist the West's Carbon Imperialism," *Financial Times*, November 26, 2015.

45. For a useful discussion of the postcolonial obsession with secrecy, albeit with a somewhat dismissive attitude to Saha, see Itty Abraham, *The Making of the Indian Atomic Bomb: Science, Secrecy and the Postcolonial State* (London: Zed Books, 1998). For a damning indictment of India's nuclear boosters and their overhyped promises, see M. V. Ramana, *The Power of Promise: Examining Nuclear Energy in India* (New Delhi: Penguin Viking, 2012).

46. Saha, "Fuel in India" (1956), *CWMS* vol. 2, 263–67. Saha's calculations were a rebuke to the much sunnier figures provided by Homi J. Bhabha, sixteen years younger than him, from a wealthy Parsi family, and the "father" of India's nuclear program.

Chapter 6

1. J. W. von Goethe, *Elective Affinities* (Oxford: Oxford University Press, 1999), 201.

2. Jonathan Coopersmith, *The Electrification of Russia, 1880–1926* (Ithaca: Cornell University Press, 1992); Heiko Haumann, *Beginn der Planwirtschaft; Elektrifizierung, Wirtschaftsplanung und gesellschaftliche Entwicklung Sowjetrusslands, 1917–1921* (Düsseldorf: Bertelsmann, 1974); Alex G. Cummins, *The Road to NEP, the State Commission for the Electrification of Russia (GOELRO): A Study in Technology, Mobilization, and Economic Planning* (PhD thesis, University of Maryland, 1988). GOELRO stands for The State Commission for the Electrification of Russia.

3. Gosplan, "Plenum Gosplana," *Planovoe Khoziaistvo* 6–7 (1923): 46.

4. Vladimir Illich Lenin, "Doklad Vserossiiskogo Centralnogo Ispolnitelnogo Komiteta i Soveta Narodnykh Kommissarov o Vneshnei i Vnutrennei Politike, 22 Dekabria," in *V.I. Lenin: Polnoe Sobranie Sochinenii* (Moscow: Izdatelstvo Politicheskoi Literaturoi, 1970), 159.

5. Coopersmith, *The Electrification of Russia, 1880–1926*, 296.

6. Iuri N. Flakserman, *Gleb Maksimilianovich Krzhizhanovskii* (Moskow: Izdatel'stvo Nauka, 1964), 245.

7. Felix Rehschuh, *Aufstieg zur Energiemacht. Der sowjetische Weg ins Erdölzeitalter, 1930er bis 1950er Jahre* (Köln: Böhlau, 2019), 12–13; Jeronim Perović, "The Soviet Union's Rise as an International Energy Power: A Short History," in *Cold War Energy: A Transnational History of Soviet Oil and Gas*, ed. Jeronim Perović (Cham: Palgrave MacMillan, 2017), 1–43.

8. Oscar Sanchez-Sibony, *Red Globalization: The Political Economy of the Soviet Cold War from Stalin to Khrushchev* (New York: Cambridge University Press, 2014), 184–92; Per Högselius, *Red Gas: Russia and the Origins of European Energy Dependence* (New York: Palgrave Macmillan, 2013).

9. Thane Gustafson, *Crisis amid Plenty: The Politics of Soviet Energy under Brezhnev and Gorbachev* (Princeton: Princeton University Press, 1989), 35–41.

10. Thane Gustafson, *Wheel of Fortune: The Battle for Oil and Power in Russia* (Cambridge: Belknap Press of Harvard University Press, 2017), 5.

11. Antoine Acker, "A Different Story in the Anthropocene: Brazil's Post-Colonial Quest for Oil (1930–1975)," *Past & Present* 249, no. 1 (2020): 167–211; Elizabeth Chatterjee, "The Asian Anthropocene: Electricity and Fossil Developmentalism," *The Journal of Asian Studies* 79, no. 1 (2020): 3–24.

12. Saul Griffith, *Electrify: An Optimist's Playbook for Our Clean Energy Future* (Cambridge: MIT Press, 2022); Nadja Popovich and Brad Plumer, "How Electrifying Everything Became a Key Climate Solution," *The New York Times*, April 14, 2023, https://www.nytimes.com/interactive/2023/04/14/climate/electric-car-heater-everything.html, accessed April 30, 2024.

13. Timothy Mitchell, "The Resources of Economics," *Journal of Cultural Economy* 3, no. 2 (2010): 192–94.

14. Natalia Krupskaia, "O Krasine," in *Leonid Borisovich Krasin: Gody Podpolia* (Moscow: Gosudarstvennoe Izdatelstvo, 1928), 137; Vladimir Karzev, *Krzhizhanovskii* (Moscow: Molodaia Gvardiia, 1980), 46.

15. Other well-known revolutionaries from this group include Leonid Krasin and Julius Martov.

16. James D. White, "The Development of Capitalism in Russia in the Works of Marx, Danielson, Vorontsov, and Lenin," in *Class History and Class Practices in the Periphery of Capitalism*, ed. Paul Zarembka (Bingley: Emerald Publishing Limited, 2019), 3–31.

17. Karzev, *Krzhizhanovskii*, 60–62.

18. Cited in Kendall E. Bailes, *Technology and Society under Lenin and Stalin: Origins of the Soviet Technical Intelligentsia, 1917–1941* (Princeton: Princeton University Press, 1978), 49.

19. Kendall E. Bailes, "Alexei Gastev and the Soviet Controversy over Taylorism, 1918–24," *Soviet Studies* 29, no. 3 (1977): 373–94.

20. Nikolai Krementsov, *Stalinist Science* (Princeton: Princeton University Press, 1996); Bailes, *Technology and Society under Lenin and Stalin*.

21. Robert William Davies, *The Soviet Economy in Turmoil, 1929–1930* (Basingstoke: Macmillan Press, 1998).

22. Coopersmith, *The Electrification of Russia, 1880–1926*, 254–56.

23. Flakserman, *Gleb Maksimilianovich Krzhizhanovskii*.

24. To this end, the Central Statistical Administration had been working on an energy balance of the economy since 1923.

25. See also V.I. Veiz, "Das Energieproblem in der gegenwärtigen Weltwirtschaft," in *The Transactions of the Second World Power Conference*, ed. World Power Conference (Berlin: VDI-Verlag, 1930). This differs from Marx's description of industrialization, where the shift to steam is a consequence of the mechanization of the working instrument.

26. In a letter to Eduard Bernstein from 27 February 1883, Engels wrote: "The steam engine taught us to transform heat into mechanical motion, but the exploitation of electricity has opened the way to transforming all forms of energy—heat, mechanical motion, electricity, magnetism, light—one into the other and back again, as well as their exploitation in industry. The circle is complete. And Deprez's latest discovery that electric currents of very high voltage can, with a comparatively small loss of energy, be conveyed by simple telegraph wire over hitherto undreamed of distances and be harnessed at the place of destination—the thing is still in embryo—this discovery frees industry for good from virtually all local limitations, enables the harnessing of even the most remote hydraulic power and, though it may benefit the towns at the outset, will in the end inevitably prove the most powerful of levers in eliminating [*Aufhebung*] the antagonism between city and countryside. Again, it is obvious that the productive forces will thereby acquire a range such that they will, with increasing rapidity, outstrip the control of the bourgeoisie." Karl Marx and Friedrich Engels, *Marx-Engels-Werke*, Band 35, *Briefe Januar 1881–März 1883* (Berlin: Dietz, 1979), 444–45.

27. Gleb M. Krzhizhanovskii, *Sochineniia III: Socialisticheskoe Stroitel'stvo*, ed. Akademia Nauk SSSR (Russian Academy of Science) (Moscow: Gosudarstvennoe energeticheskoe izdatel'stvo, 1936), 6–8.

28. To discover something by touch.

29. Gleb M. Krzhizhanovskii, "Marks o Revolucionnom Progresse Tekhniki Pri Socializme," in *Sochineniia*, Vol. III: *Socialistichekogo Stroitelstvo* (Moscow: Akademia Nauk, 1936), 346.

30. Karl Marx, *The Poverty of Philosophy* (New York: International Publishers, 1963), 109.

31. Marx, *The Poverty of Philosophy*.

32. Coopersmith, *The Electrification of Russia, 1880–1926*, 204. See for similar ideas among Socialists worldwide Charles Proteus Steinmetz, *America and the New Epoch* (New York, London: Harper & Brothers Publishers, 1916); Fred Henderson, *The Economic Consequences of Power Production* (London: G. Allen & Unwin, Limited, 1931); Thomas P. Hughes, *Networks of Power: Electrification in Western Society, 1880–1930* (Baltimore: Johns Hopkins University Press, 1993); Vincent Lagendijk, *Electrifying Europe: The Power of Europe in the Construction of Electricity Networks* (Amsterdam: Amsterdam University Press, 2008).

33. Coopersmith, *The Electrification of Russia, 1880–1926*, 43–47, 160.

34. Jonathan Coopersmith, "Soviet Electrification: The Roads Not Taken," *IEEE Technology and Society Magazine* 12, no. 2 (1993): 13–20; Coopersmith, *The Electrification of Russia, 1880–1926*, 220–22.

35. Vladimir Ilich Lenin, *Imperialism: The Highest Stage of Capitalism* (London: Penguin Books, 2010), 13–14.

36. Kurt Heinig, "Der Weg des Elektrotrusts," *Die Neue Zeit* 2, no. 39 (1912): 479–83.

37. Karzev, *Krzhizhanovskii*, 39.

38. Anne D. Rassweiler, *The Generation of Power: The History of Dneprostroi* (Oxford: Oxford University Press, 1988), 176–79.

Chapter 7

1. For typological purposes, of the three major fossil fuels it is sensible to consider oil and natural gas (petroleum) together (and coal distinctly); since natural gas is invariably produced as a byproduct of crude oil, the optimal management of each substance cannot be considered independently of production decisions regarding the other. Pérez Alfonzo was a strong supporter of the conservation to natural gas. However, this chapter focuses solely on oil.

2. For a general view of the problem, see Dag Harald Claes, *The Politics of Oil. Controlling Resources, Governing Markets and Creating Political Conflicts* (London: Edward Elgar, 2018).

3. Robert McNally, *Crude Volatility: The History and the Future of Boom-Bust Oil Prices* (New York: Columbia University Press, 2017).

4. Luis E. Giusti, "'La Apertura': The Opening of Venezuela's Oil Industry," *Journal of International Affairs* 53, no. 1 (1999), 117–28.

5. Pierre Terzian, *OPEC: The Inside Story*, trans. Michael Pallis (London: Zed Books, 1985).

6. On the evolution of fossil fuel consumption, see Simon Pirani, *Burning Up: A Global History of Fossil Fuel Consumption* (London: Pluto Press, 2018), ch. 5.

7. See, e.g., Juan Pablo Pérez Alfonzo, *Hundiéndonos en el excremento del diablo* (Caracas: El Perro y La Rana, 2009, 1st ed. 1976), 286.

8. Juan Pablo Pérez Alfonzo, *El Pentágono Petrolero. La Política nacionalista de defensa y conservación del petróleo persigue liberar al país de la excesiva dependencia de un solo recurso no renovable* (Caracas: Ediciones Revista Política, 1967); translated in English by OPEC in 2003 under the title *The Petroleum Pentagon: The Nationalist Policy of Defense and Conservation of Petroleum Seeks to Liberate the Country from the Excessive Dependence on a Sole Non-Renewable Resource* (Vienna: OPEC, 2003).

9. For a general view of Venezuelan oil politics, see Franklin Tugwell, *The Politics of Oil in Venezuela* (Stanford: Stanford University Press, 1975).

10. Giuliano Garavini, *The Rise and Fall of OPEC in the Twentieth Century* (Oxford: Oxford University Press, 2019), 39–53.

11. On the 1943 petroleum law and the later fifty-fifty profit-sharing agreement, see Bernard Mommer, *Global Oil and the Nation State* (Oxford: Oxford University Press, 2002), 85–95.

12. Juan Pablo Pérez Alfonzo, "Abstention of Doctor Juan Pablo Pérez Alfonzo," in Rómulo Betancourt, *Venezuela's Oil*, trans. Donald Peck (London: George Allen & Unwin, 1978), 160–73. The vote on the 1943 oil law became a matter of political and historical controversy against AD and its "moderation" in the confrontation with foreign companies.

13. Terzian, *OPEC*, 69–70.

14. Erich W. Zimmermann, *Conservation in the Production of Petroleum: A Study in Industrial Control* (New Haven: Yale University Press, 1957); Roger M. Olien and Dianna Davids Olien, *Oil and Ideology: The Cultural Creation of the American Petroleum Industry* (Chapel Hill: University of North Carolina Press, 2000). The TRC also enforced an end to the "flaring" of natural gas associated with oil production in the late 1940s, greatly contributing to the expansion of its use as an energy source.

15. On the history of concession contracts, see Kate Miles, *The Origins of International Investment Law: Empire, Environment and the Safeguarding of Capital* (Cambridge: Cambridge University Press, 2013).

16. Pérez Alfonzo's 1954 exchange of cables with the president's office is reproduced in Juan Pablo Pérez Alfonzo, *Petróleo, jugo de la tierra* (Caracas: El Perro y La Rana, 2011, 1st ed. 1961).

17. Juan Pablo Pérez Alfonzo, *Petróleo y Dependencia* (Caracas: Síntesis Dos Mil, 1971).

18. The reference to Canada referred to the US's Mandatory Oil Import Program of 1959, which sought to increase imports from Canada and Mexico and later, Venezuela. Pérez Alfonzo, *Petróleo y Dependencia*, 239.

19. Pérez Alfonzo, *Petróleo y Dependencia*, 75.

20. Juan Pablo Pérez Alfonzo, *Hundiéndonos en el excremento del diablo* (Caracas: El Perro y La Rana, 2009, 1st ed. 1976), 287.

21. Juan Carlos Boué, "Opec at (More Than) Fifty: The Long Road to Baghdad, and Beyond," *Oxford Energy Forum*, no. 83 (2011), 14–19. See also Manucher Farmanfarmaian, *Blood and Oil: Memoirs of a Persian Prince* (New York: Random House, 1997).

22. The best book on the rise of nationalist oil technocrats in the Arab world remains Stephen Deguid, *Technocrats, Politics and Planning: The Formulation of Arab Oil Policy, 1957–1967,* (PhD thesis: Simon Fraser University, 1976). See also Philippe Pétriat, *Aux pays de l'or noir. Une histoire arabe du pétrole* (Paris: Gallimard, 2021).

23. Christopher R. W. Dietrich, *Oil Revolution: Anticolonial Elites, Sovereign Rights, and the Economic Culture of Decolonization* (Cambridge: Cambridge University Press, 2017).

24. The agreement is reproduced in Terzian, *OPEC,* 27–28. The concordance with *The Petroleum Pentagon* is not exact: the agreement's other two principles related to increasing the refining capacity of producing countries and pursuing greater integration of the industry.

25. Abdullah Tariki and Juan Pablo Pérez Alfonzo, "World Proration Proposals," 1959–1960, box 4Jc66, Texas Independent Producers and Royalty Owners Association Records, Dolph Briscoe Center for American History, Austin, Texas.

26. Tariki and Pérez Alfonzo, "World Proration Proposals," 1959–1960, unpaginated.

27. Tariki and Pérez Alfonzo, "World Proration Proposals," 1959–1960, unpaginated.

28. Dorothea Melcher, *OPEC: La Organización de Países Exportadores de Petróleo. La prehistoria de su fundación, 1943 a 1960* (unpublished study, 2010).

29. The minutes of the OPEC Conferences from 1961 to 1986 can be found in Giuliano Garavini Collection at NYU, https://findingaids.library.nyu.edu/nyuad/ad_mc_038/.

30. NYU Abu Dhabi Archives and Special Collections, AD.MC.038. "Agreement Concerning the Creation of the Organization of Petroleum Exporting Countries (OPEC), done in Baghdad, September 14, 1960," *United Nations Treaty Series* (1962), 248–52.

31. "Juan Pérez Alfonso, Venezuelan, Regarded as Founder of OPEC," *The New York Times,* September 4, 1979.

32. Stan Steiner, "The Man Who Invented OPEC—To Conserve Energy," *The Washington Post,* September 16, 1979.

33. IEA, *Net Zero by 2050. A Roadmap for the Global Energy Sector,* Flagship Report, May 2021.

34. Grant Smith, "Saudis Dismiss Call to End Oil Spending as 'La La Land' Fantasy," *Bloomberg,* June 1, 2021.

35. On the concentration of supply for critical minerals, see IRENA, *Geopolitics of Transitions, Critical Materials,* July 2023.

Chapter 8

1. Scholars, notably James Coleman, are reluctant to label pre-1945 political leaders in Africa as nationalists, because such movements were often a reaction to specific colonial policies, usually limited to colonial capitals, and not about political independence. James S. Coleman, "Nationalism in Tropical Africa," *American Political Science Review* 48, no. 2 (1954): 404–26. However, Thomas Hodgkin offers a dissenting

opinion—which I agree on—that we can classify as nationalist: "any organisation or group that explicitly asserts the rights, claims and aspirations of a given African society (from the level of the language-group to that of 'Pan-Africa') in opposition to European authority, whatever its institutional form and objectives." Thomas Hodgkin cited in Roger K. Tangri, "The Rise of Nationalism in Colonial Africa: The Case of Colonial Malawi," *Comparative Studies in Society and History* 10, no. 2 (1968): 142.

2. *Nigerian Chronicle*, 13 August 1909, 7–8.

3. This technical description is from the *Nigerian Chronicle*, 13 August 1909, 7–8. The records do not indicate price or where coal was obtained. However, it was likely imported as of 1909. The first commercial Nigerian coal mine opened in 1915. Carolyn A. Brown, *We Were All Slaves: African Miners, Culture, and Resistance at the Enugu Government Colliery, Nigeria* (Portsmouth: James Currey, 2003), 76, 91.

4. In British West Africa, public spending slowly increased from the late nineteenth century—especially in colonies such as Lagos that had political significance but limited mining or industrial potential to attract private capital—reaching its peak during the 1950s. L. J. Butler, *Industrialisation and the British Colonial State: West Africa 1939–1951* (London: Frank Cass, 1997).

5. Palm trees were (and still are) cultivated on a large scale by farmers in Nigeria and other West African rainforest zones. Palm oil lamps are still used in Nigeria (together with kerosene lanterns, battery-powered flashlights, and candles), especially in rural communities where access to electricity is limited. Susan M. Martin, *Palm Oil and Protest: An Economic History of the Ngwa Region, South-Eastern Nigeria, 1800–1980* (Cambridge: Cambridge University Press, 1988).

6. In the early twentieth-century "artificial light was routinely viewed as the supreme sign of 'modernity' or civilisation'" in Western societies. Chris Otter, *The Victorian Eye: A Political History of Light and Vision in Britain, 1800–1900* (Chicago: University of Chicago Press, 2008), 1; on Lagos's story, see Adewumi Damilola Adebayo, "Electricity, Agency, and Class in the Lagos Colony, c. 1860s–1914," *Past & Present* 262, no. 1 (2024), 168–206.

7. "H.M." indicates "His/Her Majesty's," affirming Britain's colonial possession of the territory. *Lagos Weekly Record*, 13 December 1902, 3.

8. Ayodeji Olukoju, "Lagos in the 19th Century," in *Oxford Research Encyclopedia of African History*, May 23, 2019, https://oxfordre.com/africanhistory/view/10.1093/acrefore/9780190277734.001.0001/acrefore-9780190277734-e-281 accessed April 30, 2024.

9. On the Bight of Benin and the trans-Atlantic slave trade, see "Slave Voyages," https://www.slavevoyages.org/, accessed December 8, 2022.

10. Robert Sydney Smith, *The Lagos Consulate, 1851–1861* (Berkeley: University of California Press, 1979).

11. *Lagos Weekly Record*, 27 July 1895, 4.

12. Lagos is still Nigeria's economic capital, although the center of national government moved to Abuja in 1991.

13. R. R. Kuczynski, *Demographic Survey of the British Colonial Empire*, vol. 1 (West Africa) (London: Oxford University Press, 1948), 543–44, 548, 576–77.

14. Lagos population estimates are notoriously unreliable. Lindsay Sawyer, "PLOTTING the Prevalent but Undertheorised Residential Areas of Lagos. Conceptualising a Process of Urbanisation through Grounded Theory and Comparison" (Dr. sc., Zurich, ETH Zurich, 2016), 84, 254–55, and in general.

15. Ayodeji Olukoju, *Infrastructural Development and Urban Facilities in Lagos, 1861–2000*, New Edition [online] (Ibadan: Institut français de recherche en Afrique, 2003), 16.

16. Muritala Monsuru Olalekan, "Research Report: Urban Livelihood in Lagos 1861–1960," *Journal of the Historical Society of Nigeria* 20 (2011): 196.

17. Jillian Du and Anjali Mahendra, "Too Many Cities Are Growing Out Rather than Up. 3 Reasons That's a Problem," *World Resources Institute Insights* (blog), January 31, 2019, https://www.wri.org/insights/too-many-cities-are-growing-out-rather-3-reasons-thats-problem accessed April 30, 2024.

18. "About Lagos," Lands Bureau—Lagos State Government, https://landsbureau.lagosstate.gov.ng/about-lagos/, accessed October 20, 2022.

19. Olukoju, "Lagos in the 19th Century."

20. Colonial Lagos was a predominantly African city. For instance, out of an estimated 73,000 people in Lagos and the greater municipal area (based on the 1911 census of Southern Nigeria), only 2,940 were non-Nigerians. Kuczynski, *Demographic Survey of the British Colonial Empire*, 612.

21. Bronwen Everill, "Goods from the Sea Countries: Material Cosmopolitanism in Atlantic West Africa," in *Commercial Cosmopolitanism? Cross-Cultural Objects, Spaces, and Institutions in the Early Modern World*, ed. Felicia Gottman (London: Routledge, 2021), 108–22.

22. The first primary school was opened in 1843 while the first secondary school was opened in 1859. J. F. Ade Ajayi, *Christian Missions in Nigeria, 1841–1891: The Making of a New Elite* (London: Longman, 1965).

23. Daniel J. Paracka Jr., *The Athens of West Africa: A History of International Education at Fourah Bay College, Freetown, Sierra Leone* (New York: Routledge, 2003).

24. Alcione M. Amos, "The Amaros and Aguda: The Afro-Brazilian Returnee Community in Nigeria in the Nineteenth Century," in *The Yoruba in Brazil, Brazilians in Yorubaland: Cultural Encounter, Resilience, and Hybridity in the Atlantic World*, ed. Niyi Afolabi and Toyin Falola (Durham: Carolina Academic Press, 2017), 65–110.

25. Kristin Mann, "A Social History of the New African Elite in Lagos Colony, 1880–1913" (PhD diss., Stanford, Stanford University, 1977), 95.

26. Kristin Mann, *Slavery and the Birth of an African City: Lagos, 1760–1900* (Bloomington: Indiana University Press, 2007); A. G. Hopkins, "Innovation in a Colonial Context: African Origins of the Nigerian Cocoa-Farming Industry, 1880–1920," in *The Imperial Impact: Studies in the Economic History of Africa and India*, ed. C. Dewey and A. G. Hopkins (London: The Athlone Press, 1978), 83–96; Patrick Cole, *Modern and Traditional Elites in the Politics of Lagos* (Cambridge: Cambridge University Press, 1975).

27. The self-determination activism of this generation of Africans is discussed in

James Coleman, *Nigeria: Background to Nationalism* (Berkeley: University of California Press, 1958), especially chs. 7–9.

28. Adebayo, "Electricity, Agency, and Class in the Lagos Colony, c.1860s–1914."

29. Newspapers were the main intellectual outlet for Western-educated Lagos Africans during the nineteenth and early twentieth centuries. Derek R. Peterson, Emma Hunter, and Stephanie Newell, eds., *African Print Cultures: Newspapers and Their Publics in the Twentieth Century* (Ann Arbor: University of Michigan Press, 2016); Fred I. A. Omu, *Press and Politics in Nigeria, 1880–1937* (London: Longman, 1978).

30. Selected opinions by Africans on electricity in Lagos during the early twentieth century include *Lagos Weekly Record*, 13 March 1900, 6; *Lagos Weekly Record*, 13 December 1902, 3; *Lagos Standard*, 31 March 1909, 4; *Lagos Standard*, 7 January 1913, n.p; *Lagos Standard*, 3 June 1914, 6.

31. It is possible other Nigerians believed in privatization, but Dove is the only surviving record to my knowledge.

32. Although Dove's ideas were Lagos-specific, his article was triggered by debates in a "sister" colony, Freetown (Sierra Leone). Freetown was established by the British as a resettlement colony for emancipated Africans, liberated Africans, and their descendants. It is often regarded as a sister colony to Lagos partly because many of its liberated residents were of Nigerian (particularly Yoruba) descent. Some residents of Lagos, including Dove, had families—and often commented on socioeconomic and political events—in Freetown and vice versa. See Jean Herskovits Kopytoff, *A Preface to Modern Nigeria: The "Sierra Leonians" in Yoruba, 1830–1890* (Madison: University of Wisconsin Press, 1965).

33. Emmanuel Ayankanmi Ayandele, *The Educated Elite in the Nigerian Society* (Ibadan: Ibadan University Press, 1974), 11 and in general.

34. Coleman, *Nigeria*, 145.

35. *Nigerian Chronicle*, 13 August 1909, 7–8. Wage estimate for unskilled African workers is from Ewout Frankema and Marlous van Waijenburg, "Data on Wages, Prices and Welfare Ratio in British Africa," African Economic History Network, https://www.aehnetwork.org/data-research/data-on-wages-prices-and-welfare-ratio-in-british-africa/, accessed 16 May 2021.

36. Two bulbs (each requiring 40 watts of electricity) operating for 12 hours: 2 x 40w x 12h = 960 wh (or 0.96 kWh) rounded up to 1 kWh.

37. Leigh Gardner also recognizes Nigeria as one of the exceptions to the "general picture . . . of financial struggle across British colonial Africa" before 1914. Leigh A. Gardner, *Taxing Colonial Africa: The Political Economy of British Imperialism* (Oxford: Oxford University Press, 2012), 32.

38. *Nigerian Chronicle*, 13 August 1909, 7–8.

39. *Nigerian Chronicle*, 13 August 1909, 7–8.

40. Assuming a conservative estimate of five people per household, there would have been about 10,660 households in 1909. Yet, a power station with a 120-kilowatt (or 120,000-watt) capacity could have powered a maximum of only three thousand

40-watt bulbs at any point in time: 120,000 watts ÷ 40 watts = 3,000. For the 1909 population estimate, see Kuczynski, *Demographic Survey of the British Colonial Empire*, 543–44, 548, 576–78.

41. Peter Kilby, "Manufacturing in Colonial Africa," in *Colonialism in Africa 1870–1960*, ed. Peter Duignan and L. H. Gann, vol. 4 (Cambridge: Cambridge University Press, 1975), 498.

42. Avner Offer, "Time Horizons as Boundaries for Market, Public and Social Enterprises," Ellen McArthur Lectures, Faculty of History, University of Cambridge, Cambridge, UK, November 2018.

43. Nigel Browne-Davies, "A Précis of Sources Relating to Genealogical Research on the Sierra Leone Krio People," *Journal of Sierra Leone Studies* 3, no. 1 (2014): 1–51.

44. "Frederick William Dove (1863–1948)," FamilySearch, https://ancestors.familysearch.org/LBWT-R3Q/frederick-william-dove-1863-1948, accessed May 17, 2021.

45. K. A. B. Jones-Quartey, "Sierra Leone's Role in the Development of Ghana (Reprinted from *Sierra Leone Studies*, New Series No. 11, December 1958)," https://www.natinpasadvantage.com/Sierra_Leone_History/Sierra_Leones_Role_in_the_Development_of_Ghana-part-1.html, accessed May 17, 2021.

46. Julie Crooks, "Alphonso Lisk-Carew: Early Photography in Sierra Leone" (PhD diss., London, SOAS, University of London, 2014), 108; The Open University, Making Britain: Discover How South Asians Shaped the Nation, 1870–1950 (database), *The Africa and Orient Review*, https://www.open.ac.uk/researchprojects/makingbritain/content/africa-and-orient-review, accessed May 17, 2021.

47. Nigel Browne-Davies, "Jewish Merchants in Sierra Leone, 1831–1934," *Journal of Sierra Leone Studies* 6, no. 1 (2017): 48.

48. Ayandele, *The Educated Elite in the Nigerian Society*, ch. 1 and in general. See also Ndubueze L. Mbah, "The Black Englishmen of Old Calabar: Freedom and Mobility in the Age of Abolition in West Africa," *Radical History Review* 2022, no. 144 (2022): 45–75.

49. A. Adu Boahen, "Politics and Nationalism in West Africa, 1919–35," in *Africa under Colonial Domination 1880–1935*, ed. A. Adu Boahen, vol. vii, *General History of Africa* (London: Heinemann Educational Books, 1985), 634; Ayandele, *The Educated Elite in the Nigerian Society*, 9–54.

50. Olúfẹ́mi Táíwò, *How Colonialism Preempted Modernity in Africa* (Bloomington: Indiana University Press, 2010), 98–127.

51. Thomas Hughes, *Networks of Power: Electrification in Western Society, 1880–1930* (Baltimore: John Hopkins University Press, 1983). See also Richard Gilbert and Edward Kahn, eds., *International Comparisons of Electricity Regulation* (Cambridge: Cambridge University Press, 1996), 1–22.

52. Renfrew Christie, *Electricity, Industry and Class in South Africa* (London: Macmillan, 1984), 27–45.

53. See, among others, Toyin Falola, *Development Planning and Decolonization in Nigeria* (Gainesville: University Press of Florida, 1996).

54. J. N. Oliphant to Chief Secretary to the Government, 25 September 1944,

"Hydro-Electric Projects," Files Assembled from the old secretariat Lagos in May 1960, Series 4, 42567, Nigerian National Archives Ibadan.

55. David Parker, *The Official History of Privatisation* (London: Routledge, 2009), 1:3.

56. Jonathan Coopersmith, *Electrification of Russia, 1880–1926* (Ithaca: Cornell University Press, 1992); Daniela Russ, this book, ch. 6.

57. See also other relevant chapters in Richard Gilbert and Edward Kahn, eds., *International Comparisons of Electricity Regulation* (Cambridge: Cambridge University Press, 1996).

58. Richard Gilbert and Edward Kahn, "Competition and Institutional Change in US Electric Power Regulation," in *International Comparisons of Electricity Regulation*, ed. Richard Gilbert and Edward Kahn (Cambridge: Cambridge University Press, 1996), 179–230; Ronald C. Tobey, *Technology as Freedom: The New Deal and the Electrical Modernization of the American Home* (Berkeley: University of California Press, 1996).

59. Seavey Clyde, "Functions of the Federal Power Commission," *Annals of the American Academy of Political and Social Science* 201, no. 1 (1939), 73–81.

60. Naomi Oreskes and Eric Conway, *The Big Myth: How American Business Taught Us to Loathe Government and Love the Free Market* (New York: Bloomsbury, 2023), ch. 2.

61. Myrna Alexander, "Reflections on the Role of the World Bank in Public Enterprises and Privatization in Africa," in *Public Enterprise at the Crossroads: Essays in Honour of V. V. Ramanadham*, ed. John Heath (London: Routledge, 1990), 199–201.

62. NERC, "Our History," accessed December 8, 2022, https://nerc.gov.ng/index.php/home/nesi/401-history.

63. Bankole Orimisan, "LCCI Urges FG to Privatise Electricity Transmission Company," *The Guardian*, October 19, 2022, https://guardian.ng/business-services/lcci-urges-fg-to-privatise-electricity-transmission-company/; "Poor Execution Bane of Power Sector Privatisation–NDPHC," *The Sun Nigeria*, November 28, 2022, https://www.sunnewsonline.com/poor-execution-bane-of-power-sector-privatisation-ndphc/; "Navigating Corruption in Africa's Power Sector," *African Business Magazine*, June 18, 2015, https://african.business/2015/06/energy-resources/navigating-corruption-in-africas-power-sector.

64. Chika Izuora, "'Nigeria Requires 30,000MW Electricity Generation to Meet Current Demand,'" *Leadership*, August 10, 2022, https://leadership.ng/nigeria-requires-30000mw-electricity-generation-to-meet-current-demand/, accessed April 30, 2024; Okechukwu Nnodim, "Grid Collapses 98 Times under Buhari amid N1.52tn Bailout," *Punch*, October 17, 2022, https://punchng.com/grid-collapses-98-times-under-buhari-amid-n1-52tn-bailout/, accessed April 30, 2024.

65. Chika Izuora, "'Nigeria Requires 30,000MW.'"

66. Simon Kolawole, "The Politics of Electricity Tariff," *TheCable*, July 5, 2020, https://www.thecable.ng/the-politics-of-electricity-tariff, accessed April 30, 2024; Simon Kolawole, "Playing Politics with Power Sector," *TheCable*, June 27, 2020, https://www.thecable.ng/playing-politics-with-power-sector, accessed April 30, 2024.

67. "Abuja Residents Decry High Electricity Tariff, Threaten Protest," *Vanguard News*, October 24, 2022, https://www.vanguardngr.com/2022/10/abuja-residents-decry-high-electricity-tariff-threaten-protest/, accessed April 30, 2024; Akinkunmi Obakeye, "Protest Rocks Ibadan Over Hike in Fuel Price, Electricity Tariff," *Channels Television*, September 8, 2020, https://www.channelstv.com/2020/09/08/protest-rocks-ibadan-over-hike-in-fuel-price-electricity-tariff/, accessed April 30, 2024.

68. Busola Aro, "Nigeria Not Ripe for Total Power Privatisation, Says Electricity Workers," *TheCable*, September 8, 2022, https://www.thecable.ng/nigeria-not-ripe-for-total-power-privatisation-says-electricity-workers, accessed April 30, 2024; OpeOluwani Akintayo, "Should Nigeria Reverse Nine-Year Power Privatisation?," *Punch*, August 16, 2022, https://punchng.com/should-nigeria-reverse-nine-year-power-privatisation/, accessed April 30, 2024.

69. Among others, Philip Akpen, "A History of Electricity and Water Supply in Selected Northern Nigerian Cities: A Comparative Study of Lokoja, Jos and Kaduna, ca. 1900–1960" (PhD diss., Kano, Bayero University, 2010); Ayodeji Olukoju, "'Never Expect Power Always': Electricity Consumers' Response to Monopoly, Corruption and Inefficient Services in Nigeria," *African Affairs* 103, no. 410 (2004): 51–71; Marcel N. Azodo Manafa, *Electricity Development in Nigeria, 1896–1972* (Yaba [Nigeria]: Raheem Publishers, 1979).

70. The main exceptions include Adebayo, "Electricity, Agency, and Class in the Lagos Colony, c.1860s–1914"; Olukoju, "'Never Expect Power Always.'"

Chapter 9

1. *Soudanese* here refers to the geographic region of the western Sudan (or Soudan in French). In 1923 much of this territory was under French colonial rule and was referred to as the French Soudan (Soudan Français). It is today the West African nation the Republic of Mali.

2. In the latter half of the twentieth century, after women had adopted faster cooking metal pots, women still frequently spent up to three hours cooking over a fire. See Jacqueline Ki-Zerbo, "Women and the Energy Crisis in the Sahel," *Unasylva* 33, no. 133 (1981): 5–10.

3. Dominique Zahan, *Antilopes du Soleil: Arts et Rites Agraires d'Afrique Noire* (Wien: Édition A. Schendl, 1980), 35–41, 109.

4. The original Bambara and French texts refer to a granary for millet but also the grain sorghum for the meal. As millet is often employed as a wide category of related but distinct grains cultivated and consumed across the region, I use the term *millet* in my discussion.

5. Historian Emily Osborn similarly centers the role of human labor in studying the energy history of transportation infrastructures and containers for the same region. See Emily Lynn Osborn, "Containers and Mobility in West Africa," *History and Anthropology* 29, no. 1 (2018): 21–31, here 27–29.

6. Vincent Bonnecase, "When Numbers Represented Poverty: The Changing Meaning of the Food Ration in French Colonial Africa," trans. Rachel Kantrowitz,

Journal of African History 59, no. 3 (2018): 463–81; Deborah Neill "Finding the 'Ideal Diet': Nutrition, Culture, and Dietary Practices in France and French Equatorial Africa, c. 1890s-1920s," *Food and Foodways* 17, no. 1 (2009): 1–28.

7. See chs. 1 and 4 in Laura Ann Twagira, *Embodied Engineering: Gendered Labor, Food Security, and Taste in Twentieth-Century Mali* (Athens: Ohio University Press, 2021), 37–40, 143–46.

8. Charles Bird (1974) cited in Patrick McNaughton, *The Mande Blacksmiths: Knowledge, Power, and Art in West Africa* (Bloomington: Indiana University Press, 1993), 15.

9. Patrick McNaughton, "The Shirts That Mande Hunters Wear," *African Arts* 15, no. 3 (1982): 54–58, 91, here 56. *Nyamaw* is the plural form of *nyama*.

10. McNaughton, *The Mande Blacksmiths*, 15–18.

11. As scholar Jacqueline Ki-Zerbo points out, women in the wider Sahel region rarely serve uncooked food. See Ki-Zerbo, "Women and the Energy Crisis in the Sahel."

12. Barbara G. Hoffman, "Power, Structure, and Mande *jeliw*," in *Status and Identity in West Africa: Nyamakalaw of Mande*, ed. David C. Conrad and Barbara E. Frank (Bloomington: Indiana University Press, 1995), 36–45, here 37–40.

13. Cara New Daggett, *The Birth of Energy: Fossil Fuels, Thermodynamics, and the Politics of Work* (Durham: Duke University Press, 2019), 19.

14. Bambara and Malinke communities share a common Mande linguistic and cultural heritage. Malinke are also sometimes identified as Maninka, Madinka, Manding, or Mandingo; Bambara is the anglicized form of Bamana.

15. Jean-Hervé Jézéquel has examined Moussa Travélé's intellectual career as a researcher and protégé of the French administrator-ethnographer Maurice Delafosse. See Jean-Hervé Jézéquel, "Les professionels Africain de la recherche dans l'état colonial tardif," *Revue d'Histoire des Sciences Humaines* 1, no. 24 (2011): 35–60; Jean-Hervé Jézéquel, "Maurice Delafosse et l'emergence d'une littérature Africaine à vocation scientifique," in *Maurice Delafosse: Entre Orientalisme et Ethnographie* (Paris: Maisonneuve et Larose, 1998).

16. Moussa Travélé, *Proverbes et contes bambaras* (Paris: Librarie Orientaliste Paul Geuthner, 1923), 46. Here I reproduce Travélé's original Bambara transcription along with his translation and short commentary in French. All translations from the French to English are by the author unless otherwise noted.

17. Laurence Porter examined a set of Wolof stories collected and translated by African school teachers in Senegal in the 1940s and found that the teachers made compromises in their final translations to accommodate foreign readers, while at the same time preserving essential elements of the stories. Laurence M. Porter, "Lost in Translation: From Orature to Literature in the West African Folktale," *Symposium* 49, no. 3 (1995): 229–39; here 229–30.

18. In 1978 Maurice Houis reviewed the collection and praised the book for its invaluable documentation. However, he lamented the form in which Travélé transcribed the Bambara text. Houis even suggests that the Bambara transcriptions were difficult to read and therefore unusable because Travélé wrote Bambara by ear and according

to French linguistic pronunciation. Certainly by 1978 standardized versions of writing in Bambara emerged, but Travélé is to be credited for his early role as one of the first scholars and ethnographers to write in Bambara. In fact, it is entirely possible to read the Bambara versions of Travélé's collected stories if one is sufficiently versed in the language. Maurice Houis, "Travélé (Moussa): Proverbes et contes bambara," *Revue français d'histoire d'outre-mer* 65, no. 238 (1978): 128.

19. Roderick McIntosh, "Social Memory in Mande," in *The Way the Wind Blows: Climate, History, and Human Action*, ed. Tainter McIntosh (New York: Columbia University Press, 2000). Harold Scheub similarly argues that oral tradition in Africa contains fragmentary episodes of historic information, as well as a window into modes of historical interpretation. See Harold Scheub, "A Review of African Oral Traditions and Literature," *African Studies Review* 28, no. 2/3 (1985): 1–72. See also Luise White, *Speaking with Vampires: Rumor and History in Colonial Africa* (Berkeley: University of California Press, 2000).

20. Mary Jo Arnoldi, "Wild Animals and Heroic Men: Visual and Verbal Arts in the 'Sogo bò' Masquerades of Mali," *Research in African Literatures* (2000): 63–75, see especially 67–70.

21. White, *Speaking with Vampires*, 8–9.

22. Mariam Thiam, "The Role of Women in Rural Development in the Segou Region of Mali," in *Women Farmers in Africa: Rural Development in Mali and the Sahel*, ed. Lucy E. Creevey (Syracuse: Syracuse University Press, 1986), 76–77.

23. Zahan, *Antilopes du Soleil*, 109, 148–51.

24. Birama Diakon, *Office du Niger et pratiques paysannes: Appropriation technologique et dynamique sociale* (Paris: l'Harmattan, 2012), 34.

25. Charles Monteil, *Contes Soudanais* (Paris: Ernest Leroux, 1905), 57–61.

26. John Cropper, "'The Sparrow Loves Millet, but Labors Not': Energy Use and Infrastructure in the Senegal Valley, 1450–1760," *History and Technology* 39, no.1 (2023): 42–64, here 47. The story "Bolde" is translated into English in David P. Gamble, "Wolof Stories from Senegambia," *Gambia Studies*, no. 10 (1987): 30–33.

27. Cropper, "'The Sparrow Loves Millet,'" 44.

28. Hamady Bocoum, "Samori's Smithies: From Craft Production to Attempted Manufacturing, or a Draft Plan for Technological Independence," *Mande Studies* (2001): 55–63, here 57–62. Samori Touré founded an empire in the late nineteenth century and is remembered both as a violent enslaver and as a valiant anti-French fighter. He is also associated with a wave of popular Islamic conversion during the same time period. See Brian J. Peterson, *Islamization from Below: The Making of Muslim Communities in Rural French Sudan, 1880–1960* (New Haven: Yale University Press, 2011).

29. Brian J. Peterson, "History, Memory and the Legacy of Samori in Southern Mali, c. 1880–1898," *The Journal of African History* 49, no. 2 (2008): 261–79.

30. Eugenia Herbert, *Iron, Gender, and Power: Rituals of Transformation in African Societies* (Bloomington: Indiana University Press, 1993).

31. See ch. 3 in Emily S. Burrill, *States of Marriage: Gender, Justice, and Rights in Colonial Mali* (Athens: Ohio University Press, 2015), 79–105.

32. Diakon, *Office du Niger*, 29.

33. Zahan, *Antilopes du Soleil*, 20–21, 31.

34. Diakon, *Office du Niger*, 34–41, 43–44; on gendering energy technologies, see Turnbull, this book, ch. 11.

35. Twagira, *Embodied Engineering*, 142–56.

36. The "for life" (*ka balo*) field is tended collectively by households for consumption while "for money" (*ka wari nyini*) fields are more often individual endeavors. Stephen Wooten, *The Art of Livelihood: Creating Expressive Agri-Culture in Rural Mali* (Durham: Carolina Academic Press, 2009), 65–66.

Chapter 10

1. William James, *The Energies of Men* (New York: Moffat, Yard & Company, 1914), 6.

2. There is an extensive literature on the life and work of William James, including a specialist journal dedicated to *William James Studies*.

3. See Sam Halliday, *Science and Technology in the Age of Hawthorne, Melville, Twain, and James: Thinking and Writing Electricity* (New York: Palgrave Macmillan, 2007); Jennifer Lieberman, *Power Lines: Electricity in American Life and Letters, 1882–1952* (Cambridge: Cambridge University Press, 2017).

4. I am heavily indebted to Sergio Franzese's interpretation of William James energy ethics developed in *The Ethics of Energy: William James's Moral Philosophy in Focus* (Boston: De Gruyter, 2013). Franzese demonstrates how James lodged energy at the heart of his pragmatist philosophy.

5. Robert D. Richards, *William James, In the Maelstrom of American Modernism* (New York: Houghton Mifflin Harcourt, 2013), 84.

6. Richards, *William James*, 84.

7. This lecture was published in *The Philosophical Review* in 1907 and is the version quoted from in this chapter. William James, "The Energies of Men," *The Philosophical Review* 16, no. 1 (1907), 1–20.

8. James, "The Energies of Men," 3. Francesca Bordogna shows how James transgressed disciplinary conventions in this speech. See Francesca Bordogna, *William James at the Boundaries: Philosophy, Science and the Geography of Knowledge* (Chicago: University of Chicago Press, 2002).

9. James, "The Energies of Men," 2.

10. James, "The Energies of Men," 3.

11. The saturation of energy in the American imaginary during the Progressive Era culture is explored further in Rebecca Wright, *Moral Energy in America: From the Progressive Era to the Atomic Bomb* (Baltimore: Johns Hopkins University Press, 2025).

12. Although focussed predominantly on Europe, Anson Rabinbach's *The Human Motor: Energy, Fatigue, and the Origins of Modernity* (Berkeley: University of California Press, 2016) demonstrates how energy was incorporated into the science of work.

13. See David Nye, *Electrifying America: Social Meanings of a New Technology, 1880–1940* (Cambridge: MIT Press, 1990).

14. See Sarah Watts, *Rough Rider in the White House and the Politics of Desire* (Chicago: University of Chicago Press), 2003.

15. Carolyn de la Peña, *The Body Electric: How Strange Machines Built the Modern American* (New York: New York University Press, 2005), 28–29.

16. George Beard, *American Nervousness, Its Causes and Consequences: A Supplement to Nervous Exhaustion (Neurasthenia)* (New York: G. P. Putnam's Sons, 1881); David G. Schuster, *Neurasthenic Nation: America's Search for Health, Happiness, and Comfort, 1869–1920* (New Brunswick: Rutgers University Press, 2011)

17. De la Peña, *The Body Electric*, 26.

18. James, "The Energies of Men," 2.

19. James, "The Energies of Men," 18.

20. Henry Adams, *A Letter to American Teachers of History* (Washington, 1910).

21. William James to Henry Adams, June 17, 1910, in *The Letters of William James*, Vol. 2, ed. Henry James (Frankfurt: Outlook Press, 2018), 245.

22. James to Adams, June 17, 1910, 245.

23. James, "The Energies of Men," 9.

24. Franzese, *The Ethics of Energy*, 185.

25. James, "The Energies of Men," 19.

26. James, *The Energies of Men* (1914), 10–11.

27. James, "The Energies of Men," 12.

28. William Cotkin, *William James: Public Philosopher* (Baltimore and London: Johns Hopkins University Press, 1990), 114.

29. Franzese, *The Ethics of Energy*, 6.

30. He first raised this metaphor in William James, *The Varieties of Religious Experience: A Study in Human Nature* (London: Longmans, Green and Company, 1902), 367.

31. William James, "The Moral Equivalent of War," in *The Writing of William James: A Comprehensive Edition*, ed. John J. McDermott (Chicago: University of Chicago Press, 1977), 669.

32. I discuss this further in Wright, *Moral Energy in America* (forthcoming). Paul Boyer, *Urban Masses and Moral Order in America* (Cambridge: Harvard University Press, 1978).

33. Harriet Hickox Heller, "The Playground as a Phase of Social Reform," in *Proceedings of the Second Annual Playground Congress*, New York, September 8–12 (New York: Playground Association of America, 1908), 179.

34. Charlotte Perkins Gilman, *Women and Economics—A Study of the Economic Relations between Men and Women as a Factor in Social Evolution* (Boston: Small, Maynard & Company, 1898).

35. Charlotte Perkins Gilman, *The Home: Its Work and Influence* (New York: Charlton Company, 1903).

Chapter 11

1. Dominic Boyer, "Energopolitics and the Anthropology of Energy," *Anthropology News* (May 2011): 1–7; Tristan Loloum, Simone Abram, and Nathalie Ortar, *Eth-*

nographies of Power: A Political Anthropology of Power (London: Berghahn, 2021), 1–2, inter alia.

2. Leslie White argued for a hierarchised theory of energy-determined development in which cultural evolution (P) was described as a function of per-capita energy use (E) and the efficiency of this use (F), leading to the supposedly explanatory formula $P = EF$. Leslie White, "Energy and the Evolution of Culture," *American Anthropologist* 45, no. 3 (1943): 338; Richard Barrett, "The Paradoxical Anthropology of Leslie White," *American Anthropologist* 91, no. 4 (1989): 988.

3. Amid the differentially allocated energy crises of the 1970s, Adams called for an energy anthropology that addressed "mankind as a whole." Richard Adams, "Man, Energy, and Anthropology: I Can Feel the Heat, But Where's the Light?," *American Anthropologist* 80, no. 2 (1978): 297–309; on Rappaport, see Brian Hoey and Tom Fricke, "'From Sweet Potatoes to God Almighty': Roy Rappaport on Being a Hedgehog," *American Ethnologist* 34, no. 3 (2007): 587.

4. Laura Nader, "Energy Choices: Humanity's Role in Changing the Face of the Earth," *Social Analysis* 56, no. 3 (2012): 109.

5. Simon Schaffer, *From Physics to Anthropology and Back Again* (Cambridge: Prickly Pear Press 1994), 7.

6. George Stocking, "From Physics to Ethnology: Franz Boas' Arctic Expedition as a Problem in the Historiography of the Behavioral Sciences," *Journal of the History of the Behavioral Sciences* 1, no. 1 (1965): 53–66.

7. Richard Handler, "Boasian Anthropology and the Critique of American Culture," *American Quarterly* 42, no. 2 (1990): 255–58.

8. Charles King, *The Reinvention of Humanity: How a Circle of Renegade Anthropologists Remade Race Sex and Gender* (London: Vintage Books, 2020), 38–57.

9. White lamented Boas'a rejection of evolutionary theory in favour of "planless hodge-podge-ism". White, "Energy and the Evolution of Culture," 355; In White's reviving of energo-evolutionism there are parallels to the reemergence of climatic determinism. Mike Hulme, "Reducing the Future to Climate: A Story of Climate Determinism and Reductionism," *Osiris* 26, no. 1 (2011): 245–66.

10. The idea of three stages of energy anthropology can be found in Dominic Boyer, "Energopower: An Introduction," *Anthropological Quarterly* 87, no. 2 (2014): 312. He and Cymene Howe, appreciative of Nader, have done much to advance the subdiscipline.

11. Laura Nader, "Barriers to Thinking New about Energy," *Physics Today* 34, no. 2 (1981): 9–104.

12. David Price, "Contextualizing Laura Nader," in *Introduction to Laura Nader: A Life of Teaching, Investigation, Scholarship and Scope* (Berkeley: University of California Press, 2013), 1–3, 61.

13. Laura Nader, "The Phantom Factor: Impact of the Cold War on Anthropology," in *The Cold War and the University: Toward an Intellectual History of the Postwar Years*, ed. Noam Chomsky, Ira Katznelson, R.C. Lewontin, David Montgomery, Laura Nader, Richard Ohmann, Ray Siever, Immanuel Wallerstein, and Howard Zinn (New York: New Press, 199), 128.

14. Laura Nader, *Harmony Ideology: Justice and Control in a Zapotec Mountain Village* (Stanford: Stanford University Press, 2013).

15. Margaret Rossiter, *Women Scientists in America: Before Affirmative Action, 1940–1972* (Baltimore: Johns Hopkins, 1982): 142

16. Price, "Contextualising Laura Nader," 75.

17. Kelly Moore, *Disrupting Science: Social Movements, American Scientists, and the Politics of the Military, 1945–1975* (Princeton: Princeton University Press, 2008), 50.

18. William Rorabaugh, *Berkeley at War: The 1960s* (Oxford: Oxford University Press, 1989), 85, 122.

19. Nader, "The Phantom Factor," 128–29.

20. Laura Nader, "Up the Anthropologist—Perspectives Gained from Studying Up," in *Reinventing Anthropology*, ed. Dell Hymes (New York: Vintage Books, 1974), 5.

21. Nader, "The Phantom Factor," 128.

22. Nader, "The Phantom Factor," 123–24.

23. Nader, "The Phantom Factor," 108–9; Susan Trencher, *Mirrored Images: American Anthropology and American Culture, 1960–1980* (Westport: Bergin & Garvey, 2000), 134–136.

24. Nader, "The Phantom Factor," 113; David Price "Cold War Anthropology: Collaborators and Victims of the National Security State," *Identities: Global Studies in Culture and Power* 4, no. 3–4 (1998): 389–430.

25. Clyde Kluckhohn, *Mirror for Man: The Relation of Anthropology to Modern Life* (New York: McGraw Hill, 1949), 261.

26. Mauro Barbosa de Almeida, "Symmetry and Entropy: Mathematical Metaphors in the Work of Lévi-Strauss," *Current Anthropology* 31, no. 4 (1990): 367–385.

27. Akin to the later notion of the technosphere, Lévi-Strauss wrote, "Civilization, taken as a whole, can be described as an extraordinarily complex mechanism, which we might be tempted to see as offering an opportunity of survival for the human world, if its function were not to produce what physicists call entropy." See Claude Lévi-Strauss, *Tristes Tropiques*, trans. John and Doreen Weightman (London: Penguin, 1973), 543.

28. Philip Descola, *In the Society of Nature: A Native Ecology in Amazonia* (Cambridge: Cambridge University Press, 1994), 314–16.

29. Hugh Gusterson, "Studying Up Revisited," *Political and Legal Anthropology Review* 20, no. 1 (1997): 117.

30. Roberto Gonzalez, Laura Nader, and C. Jay Ou, "Between Two Poles: Bronislaw Malinowski, Ludwig Fleck, and the Anthropology of Science," *Current Anthropology* 36, no. 5 (1995): 865–69.

31. John Gibbons and William Chandler, *Energy: The Conservation Revolution* (New York: Plenum, 1981), 31–32.

32. Price, "Contextualizing Laura Nader," 132.

33. Harvey Brooks and Jack Hollander, "United States Energy Alternatives to 2010 and Beyond: The CONAES Study," *Annual Review of Energy* 4, no. 1 (1979): 1–70.

34. Harvey Brooks, "Perspectives on the Energy Problem," *Proceedings of the American Philosophical Society* 125, no. 4 (1981): 245–50; Brooks and Hollander, "United States Energy Alternatives," 67.

35. Laura Nader and Stephen Beckerman, "Energy as It Relates to the Quality and Style of Life," *Annual Review of Energy* 3, no. 1 (1978): 19.

36. Nader and Beckerman, "Energy."

37. Nader and Beckerman, "Energy," 10.

38. Nader and Beckerman, "Energy," 25.

39. Price, "Contextualizing Laura Nader," 24; Laura Nader, *Energy Choices in a Democratic Society: The Report of the Consumption, Location, and Occupational Patterns Resource Group, Panel of the Committee on Nuclear and Alternative Energy Systems* (Washington, DC: National Academy of Sciences, 1980).

40. Nader, *Energy Choices*, 8.

41. Nader, *Energy Choices*, 56, 84.

42. Nader and Beckerman, "Energy," 2.

43. Nader and Beckerman, "Energy," 2. Leslie White reviewed Clark and Childe and praised their having taken a stand against the "reactionary doctrine of anticultural evolutionism." See Leslie White, "Review: *From Savagery to Civilization* by Grahame Clark [and] *History* by V. Gordon Childe [. . .]," *Antiquity* 22, no. 88 (1948): 216–17.

44. Nader and Beckerman, "Energy," 6.

45. Howard Odum and Richard Pinkerton, "Time's Speed Regulator: The Optimum Efficiency for Maximum Output in Physical and Biological Systems," *American Scientist* 43, no. 2 (1953): 331.

46. Price, "Contextualizing Laura Nader," 73.

47. Laura Nader, "The Three-Cornered Constellation: Magic, Science, and Religion Revisited," in *Naked Science: Anthropological Inquiry into Boundaries, Power, and Knowledge*, ed. Laura Nader (New York: Routledge, 2014), 262.

48. Nader and Beckerman, "Energy," 6.

49. A full account of the talk is given in Laura Nader, *Contrarian Anthropology: The Unwritten Rules of Academia* (Oxford: Berghahn, 2018), 33–44.

50. Laura Nader, "Controlling Processes: Tracing the Dynamic Components of Power—Sidney W. Mintz Lecture for 1995," *Current Anthropology* 38, no. 77 (1997): 733.

51. Laura Nader, "Ethnography as Theory," *HAU: Journal of Ethnographic Theory* 1, no. 1 (2011), 217.

52. Nader, *Naked Science*, 266.

53. Nader, *Naked Science*, 272.

54. Nader, *Naked Science*, 273.

55. Laura Nader and Stephen Beckerman, "Questions for an Ethnography of Energy," *Anthropology Resource Center Newsletter* 2, no. 1 (1978): 2.

56. Victor McFarland, "The United States and the Oil Price Collapse of the 1980s," in *Counter-Shock: The Oil Counter-Revolution of the 1980s*, ed. Duccio Basosi et al. (Oxford: IB Taurus, 2019): 259–73.

57. Simon Bromley, *American Hegemony and World Oil: The Industry, the State System and the World Economy* (Polity: Cambridge: Polity, 1991), 207.

58. Regina Axelrod and Hugh Wilson, "Reagan's Concept of Federalism and Nuclear Power: Shoreham—A Case of Conflict," *Energy Policy* 19, no. 9 (1991): 841–48.

59. Laura Nader, "Harmony Coerced Is Freedom Denied," *Chronicle of Higher Education*, July 13, 2001, https://www.chronicle.com/article/harmony-coerced-is-freedom-denied/.

60. Laura Nader, "Energy Choices: Humanity's Role in Changing the Face of the Earth," *Social Analysis* 56, no. 3 (2012): 108–116; Andrea Behrends, Stephen Reyna, and Gunther Schlee, *Crude Domination: An Anthropology of Oil* (New York: Berghahn, 2007).

61. Nader, "Controlling Processes," 734–35.

62. Jaia Syvitski et al., "Extraordinary Human Energy Consumption and Resultant Geological Impacts Beginning around 1950 CE Initiated the Proposed Anthropocene Epoch," *Communications Earth and Environment* 1, no. 32 (2020): 1–13.

63. Nader, "Energy Choices," 108.

Chapter 12

1. I would like to acknowledge the useful, kind, and fun help I received with this chapter from Liu Xinyue, Gregory Vettese, Cameron Hu, Daniela Russ, and Thomas Turnbull. Of course, only I am responsible for remaining errors and infelicitous phrasing.

2. Bernard Dixon, "In Praise of Prophets," *New Scientist and Science Journal* 51, no. 769 (1971): 606.

3. Thomas Robertson, *The Malthusian Moment: Global Population Growth and the Birth of American Environmentalism* (New Brunswick: Rutgers University Press, 2012).

4. Paul Ehrlich and Peter H. Raven, "Butterflies and Plants: A Study in Coevolution," *Evolution* 18, no. 4 (1964): 586–608.

5. Thomas Malthus, *An Essay on the Principle of Population*, 2nd ed. (London: J. Johnson 1803 [1798]), 14.

6. For a clarifying discussion on neoliberal epistemology, see Philip Mirowski, *Never Let a Serious Crisis Go to Waste: How Neoliberalism Survived the Financial Meltdown* (London: Verso, 2014), 54. Two important neoliberal essays in this vein include Friedrich Hayek, "Economics and Knowledge," *Economica* 4, no. 13 (1937): 33–54; and Hayek, "The Use of Knowledge in Society," *American Economic Review* 35, no. 4 (1945): 519–30.

7. Paul Sabin, *The Bet: Paul Ehrlich, Julian Simon, and Our Gamble over Earth's Future* (New Haven: Yale University Press, 2013).

8. Letter from Paul Ehrlich to Julian Simon, 1 October 1980; and letter from Julian Simon to Paul Ehrlich, 17 October 1980, box 21, folder Paul Ehrlich, Correspondence with and about (6660), 1971–72, 1974, 1977–78, 1980–81, 1985, Julian Simon Papers, University of Illinois, Urbana-Champaign Archive.

9. John Tierney, "Betting on the Planet," *New York Times*, December 2, 1990.

10. Julian Simon, *The Ultimate Resource* (Princeton: Princeton University Press, 1981), 348.

11. Simon, *The Ultimate Resource*, 49.

12. Simon, *The Ultimate Resource*, 91–92.

13. Sabin, for instance, does not investigate how resourceship belongs to the broader neoliberal school of thought.

14. This biographical section is largely drawn from Sabin, *The Bet*, 68–78.

15. For an overview of Zimmermann's work, see Thomas Turnbull, "Toward Histories of Saving Energy: Erich Walter Zimmermann and the Struggle against 'One-Sided Materialistic Determinism,'" *Journal of Energy History/Revue d'histoire de l'énergie (JEHRHE)* 4 (2020): 1–21.

16. Zimmermann, *World Resources and Industries: A Functional Appraisal of the Availability of Agricultural and Industrial Resources* (New York and London: Harper & Brothers, 1933), 765.

17. Harold Barnett and Chandler Morse, *Scarcity and Growth: The Economics of Natural Resources Availability* (Baltimore: Resources for the Future, 1963).

18. Cited in Sabin, *The Bet*, 78.

19. Zimmermann, *World Resources and Industries*, 71.

20. Robert Solow, "The Economics of Resources or the Resources of Economics," *American Economic Review* 64, no. 2 (1974): 7.

21. Simon, *The Ultimate Resource*, 99.

22. Friedrich Hayek, *The Constitution of Liberty* (Chicago: University of Chicago Press, 1960), 493.

23. Hayek, *The Constitution of Liberty*, 369–70.

24. Friedrich Hayek, "The Pretence of Knowledge," *Swedish Journal of Economics* 77, no. 4 (1975): 440.

25. Letter from Friedrich Hayek to Julian Simon, 22 March 1982, box 19, folder "Economic Consequences of Population Change in Industrialized Countries," Julian Simon Papers, University of Illinois, Urbana-Champaign Archive.

26. Julian Simon, "Resources, Population, Environment: An Oversupply of False Bad News," *Science* 208, no. 4451 (1980): 1432.

27. Simon, "Resources, Population, Environment," 1432.

28. Simon, "Resources, Population, Environment," 1435.

29. Simon, "Resources, Population, Environment," 1436.

30. Simon, "Resources, Population, Environment," 1435.

31. Simon, "Resources, Population, Environment," 1436.

32. Simon, "Resources, Population, Environment," 1435.

33. Simon, "Resources, Population, Environment," 1436.

34. Weinberg and Hammond, respectively, helped create the first nuclear reactor for the Manhattan Project and worked at the Los Alamos National Laboratory. Alvin Weinberg and Philip Hammond, "Limits to the Use of Energy: The Limit to Population Set by Energy Is Extremely Large, Provided That the Breeder Reactor Is Developed or That Controlled Fusion Becomes Feasible," *American Scientist* 58, no. 4 (1970): 412–18.

35. Simon, *The Ultimate Resource*, 120–22.

36. Milton Friedman, "The Energy Boondoggle," *Newsweek*, July 30, 1979.

37. Julian Simon, *The Ultimate Resource II* (Princeton: Princeton University Press, 1998), 179.

38. Simon, "Resources, Population, Environment," 1435.

39. Simon, "Resources, Population, Environment," 1436.

40. John Holdren, Paul Ehrlich, Anne Ehrlich, and John Harte, "Bad News: Is It True?," *Science* 210, no. 4476 (1980): 1296–301.

41. Simon, "Resources, Population, Environment," 1435.

42. Holdren, Ehrlich, Ehrlich, and Harte, "Bad News: Is It True?," 1299.

43. Charles Petit, 'Two Stanford Professors Offer to Bet Optimistic Economist / Ehrlich, Schneider Say Life Getting Worse, not Better' *SF Gate*, 18 May 1995

44. Julian Simon, 'Earth's Doomsayers Are Wrong,' *SF Gate*, 12 May 1995.

45. Rosie Mestel, "'Doomsters' Take on Global Bet," *New Scientist*, 3 June 1995.

46. Robert E. Black, Lindsay H. Allen, Zulfiqar A. Bhutta, Laura E. Caulfield, Mercedes de Onis, Majid Ezzati, Colin Mathers, and Juan Rivera, "Maternal and Child Undernutrition: Global and Regional Exposures and Health Consequences," *Lancet* 371, no. 9608 (2008): 243.

47. Ehrlich himself published an authoritative study on the topic in 2015, where he and his collaborators estimated that the pace of extinctions occurs a hundred times faster than the natural "background rate." Gerardo Ceballos, Paul R. Ehrlich, and Rodolfo Dirzo, "Biological Annihilation via the Ongoing Sixth Mass Extinction Signaled by Vertebrate Population Losses and Declines," *PNAS* 114, no. 30 (2017): E6089–96.

48. Ross Emmett and Jesse Grabowski, "Better Lucky than Good: The Simon-Ehrlich Bet through the Lens of Financial Economics," *Ecological Economics* 193 (2022): 14.

49. Marion King Hubbert, *Nuclear Energy and the Fossil Fuels* 95 (Houston: Shell Development Company, Exploration and Production Research Division, 1956).

50. The chief scientific advisor to the UK government observed that "[the oil market's] behaviour up to 2005 was attributed to normal elastic supply-demand factors, but crude oil then plateaued, with the rapid price rise clearly attributable to demand exceeding conventional supply." Oliver Inderwildi and David King, "Energy Shift: Decline of Easy Oil and Restructuring of Geo-Politics," *Frontiers in Energy* 10, no. 3 (2016): 260–67.

51. Robert L. Bradley, "Resourceship: An Austrian Theory of Mineral Resources," *Review of Austrian Economics* 20 (2007): 63–90.

52. John Tierney, "Economic Optimism? Yes, I'll Take That Bet," *New York Times*, December 27, 2010.

53. As Ehrlich once quipped, "The one thing we'll never run out of is imbeciles" who are the true "ultimate resource." Tierney, "Betting on the Planet."

54. Nathalie Berta, "A Note on John Dales and the Early History of Emissions Trading: Mixing Standards and Markets for Rights," *Cahiers d'économie politique* 1 (2021): 61–84.

55. Detlef P. van Vuuren et al., "Alternative Pathways to the 1.5°C Target Reduce the Need for Negative Emission Technologies," *Nature Climate Change* 8, no. 5 (2018): 391–97.

56. "Despite frequent assertions starting in the 1970s of fundamental 'limits to growth,' there is still remarkably little evidence that human population and eco-

nomic expansion will outstrip the capacity to grow food or procure critical material resources in the foreseeable future." Ted Nordhaus, Michael Shellenberger et al. "An Ecomodernist Manifesto," ecomodernism.org (2015).

57. This section is much influenced by Tatjana Söding, "Existenzielles Risiko," *Taz*, September 9, 2023.

58. U.S. Congress, *Energy Reorganization Act of 1973: Hearings, Ninety-third Congress, First Session, on H.R. 11510* (Washington, DC: U.S. Government Printing Office, 1973), 248.

59. Simon, "Resources, Population, Environment," 1435.

Index

Abdel Nasser, Gamal, 133
Acción Democrática (AD), Venezuela, 128
Achuar people, 187
Adams, Henry, 4–5, 17, 170; entropy and civilizational decline, 173, 227; and William James, 172–73
Adams, Richard, 183
Adebayo, Damilola 15, 226
African Nationalists, 147, 151
Alberta, Canada, 208
alcohol, 22, 23, 24, 27, 28, 33; taxation on in Lagos, 147
Allahabad, University of, 87, 91
Allgemeine Elektrizitätsgesellschaft (AEG), 110, 151
Almeida, Oziel de, 34
alternative energy, 7, 12, 13, 17, 22, 27, 30, 31, 188–89, 227; in China, 55, 63; futures, 228
Ambedkar, Bhimrao Ramji, 91
American Anthropological Association, 186
American Chemical Society, 195
American Economic Association, 205
American Petroleum Institute, 211
Anglo-Iranian Oil Company, 142
Anglo-Persian Oil Company, 142
animal energy, 165
Anstey, Vera, 99
anthropocene, 5, 105, 212, 224, 227; energy and, 6, 192

anthropology, 5, 17, 183; anti-Boasian energeticism, 184; applied, 200–201; and energy, 183; entropology, 187; mainstream, 183
Apartheid South Africa, 208
Appalachian coal miners, 191
Arab Petroleum Congress, 133
Arabian American Oil Company (ARAMCO), 133
artificial intelligence, 77, 213
atomic communalism, 111
Australia, 42, 48, 214, 223

back-stop fuels, 205, 207, 212
Baffin Island, 184
Bambara, 16, 160–61, 162, 167; Bambara storytelling, 162
Barak, On, 10, 13
Barnett, Harold, 205, 217
battleship fuel, 36, 37
Beckerman, Stephen, 188
Beilby, George, 52
Berkeley, California, 17; campus protests, 185; University of California, Berkeley, 185–86
Berndt, Ernst 72
Betancourt, Rómulo, 127–28, 129
Bezos, Jeff, 213
big three, oil majors, 125
bin Salman, Abdulaziz, 135
biosphere, 212
Boas, Franz, 184, 192

274 Index

Boasian anthropology, 184, 189
Bolshevik, 104, 116
Bonneuil, Christophe, 11
Boserup, Ester, 205
Boulding, Kenneth, 213
Boulton, Matthew, 10
Brazil: Aeronautical Technology Centre (Centro de Tecnologia Aeronautica [CTA]), 25; Brazil's Atlantic Forest, 26; Brazilian Arabia, 12; Brazilian Cerrado, 27, economic miracle, 23; Ministry of Agriculture Mines and Combustion Experimental Station, 23; Pastoral Land Commission, 27; relation to Iran, 24; Ribeirão Preto, 26; Rural Worker Statute (Estatuto do Trabalhador Rural, ETR), 27; 24; São Paulo
Breakthrough Institute, 212
BRICs: 2, 11
British East India Company, 8
British Navy, 8, 10, 147
British Petroleum. *See* Anglo-Persian Oil Company
British Royal Commission, of 1871, 51
British Thermal Units (BTU), 76
Brookings Institution, 74–75, 77
Brookings, Robert, 74
Bulletin of the Geological Survey of China, 56, 58, 64, 67, 68
Burma, 9; indigenous oil industry, 94

Cadman, John, 41 47
Calcutta, 91, 92; Rotary Club, 88, 95; University of Calcutta, 86, 87
Caldwell, John, 215
calories, 36; and colonialism, 158
Canada, power use in, 97; tar sands in, 211
canons, scholarly, 19; a more representative canon, 223; volatile canon, 18–19; Western-centered canon, 223
capitalism, 4, 12, 14, 18, 131, 211; age of 108, 116, 123; carboniferous capitalism, 96; contradiction, 109–10; and endless growth, 204, 207; expansion of, 63; industrial capitalism, 4, 113; intergalactic capitalism, 213; ;monopoly capitalism, 119; socialism vs., 14, 93, 109, 119, 226
Caracas, Venezuela, 125, 129, 132, 134, 135
carbon space, 94
carbon technocracy, 12
Carnegie, Andrew, 74
Central Intelligence Agency (CIA), 186
central planning: in India, 91; in the Soviet Union, 111, 225
Central University of Venezuela, 125
charcoal, 8, 83
Chatterjee, Elizabeth, 14, 225
Chicago, University of, 204–205
Childe, Gordon Vere, 189
China, 2, 9, 10, 22; air pollution, 62; Chinese geology, 64; contemporary coal use, 55; Geological Survey of, 56; geologists, 59; geology, 60–61; inland rivers, 8; prodigious coal use, 225; Revolutionary China, 18; metallurgy, 65; mining cartography, 69; provinces: Anhui, Zhejiang, Shanxi, 58; Jiangxi, 62; Kaiping (today Hebei), 59; renewable capacity, 55; state support for mining, 65
Christian Science, 179–80
Clark, Graham, 189
Clemenceau, Georges, 36
climate change, 1, 2, 5, 31, 63, 78; and decoupling, 71; India and, 94; and limits to growth, 62; mitigation, 135; Nader and, 191; neoliberalism and, 212; OPEC and, 135
coal, 4, 8–11, 35–39, 41–44, 46–54, 55–64, 66, 92–93, 97, 100, 104, 145–46; as ballast, 10; in China, 55–64; coal-driven steamships, 9; *The Coal Question*, 42; coaling stations, 10–11; coal-to-oil synthesis, 208; and colonies, 9; com-

Index 275

bustion, 5, 49–50; development and, 12; and hegemony, 42; in India, 92–93, 97, 100; industrialism and, 4, 8, 9; in Japan, 46, 50; in Lagos, 145–46; and oil, 10; seams, 8; and slavery, 9–10; space and time and, 8; in Soviet Union, 104. *See also* Britain, China, India, Japan, Soviet Union, United States
Coal Question, The (Jevons), 42, 50, 51
coffee, 24, 134
colonialism, 85; anti-colonialism, 11, 14; *diezmo colonial*, 131; energy colonialism, 3; environmental colonialism, 94; fossil-fueled colonialism, 11; French colonialism, 16, 158; green colonialism, 2; IPCC and colonialism, 2; modernity and, 147neocolonialism, 63; and per capita energy use, 93; postcolonial critics, 84; space colonization, 204
Columbia University, 170
CONAES, Committee on Nuclear and Alternative Energy Systems, US National Academy of Science, 188, 190, 194–95; as fieldwork, 197; and soft energy, 198
concessions, oil, 127–28, 131, 135, 137, 138–39, 140–41
conservation, 12–13, 15, 40, 52–53, 76, 105, 124–26, 129–30, 132–37, 139–40, 171, 226; anti-colonial, 15, 226; conservation movement, United States, 72, 171; electricity and, 105; in Japan, 40, 52–53; oil and, 124–26, 129–30, 132–37, 139–40; policies, United States, 12, 13, 52, 76; soil conservation, 92, 101. *See also* Pérez Alfonzo; Texas Railroad Commission; Venezuela
Consumer Price Index, 217
conversion of energy, 4, 179
cooking pots, 165, 218
Coordinating Commission for the Conservation and Commerce of Hydrocarbons (CCCH), 129, 140; and OPEC, 140; origins of, 139–40; and the Texas Railroad Commission, 130
copper, 207; lodes, 219
cornucopianism, 18, 163, 203
Cornwall, England, 8
cosmopolitanism, 3, 7
COVID-19, 64, 224
Cropper, John, 163
cultural anthropology, 184

Daggett, Cara, 160
Dalit, caste, 14
Damodar Valley Corporation, 93
Danielson, Nikolai, 114
Daugherty, Carroll, 75, 80
decarbonization, 105, 175
decoupling, 71–72, 212, 213
Department of Energy (US), 198
Descola, Philippe, 187
development: developmental energies, 12, 224; historical stages of industrialization, 114. *See also* energy
diesel engines, 11
diet, of colonial subjects, 158
Dnieper Dam, 102
Dobson, Michael, 15, 226
Dove, Fredrick William, 15; life of, 150
Dutch East India Company, 131

Eaglin, Jennifer, 12, 225
Earth Day, 210
East Asia, 10
Easterlin, Richard, 205
Ebute Metta, 155
ecological optimism, 93
Ecologist, The (magazine), 202
ecomodernism, 212
Edison, Thomas 22
Effective Altruism movement, 212
Egaña, Manuel 127
Egypt, 10

Ehrlich, Paul, 202–203; the bet, 203
Einstein, Albert, 86
electric light, in Lagos, 146
Electric Power Research Institute (EPRI), 198
electricity, 15, 22, 24, 30, 46, 73, 85, 88, 91–93, 95–98, 100–101, 105, 108–109, 113, 115, 116, 121, 123, 144–54; in Africa, 144–54; age of, 88, 116, 123; and collectivization, 109; and colonialism in India, 91; and colonialism in Lagos, 149; Friedrich Engels on, 108, 115, 252; and growth, 73; in India, 85, 88, 91–93, 95–98, 100–101; as a rationalizing agent, 108; in Russia, 105, 108–109, 113, 115, 116, 121, 123. *See also* mental energy and moral energy
emancipatory energies, 13, 225
Energeía, 16
energetic optimization, 109
energetics, 116
energy, 3–5; against the energy concept, 16; and anthropocene, 193; and anthropology, 183–85; and conservation, 135; determinism, 5, 17, 76–77, 85; and economic growth, 75–76; economics, 73; "energy problem," 191; and growth, 188–89; history, 3–4; human-produced energy, 6; indeterminism, 172, 175, 188; and matter, 107; parochial, 226; poverty, 3, 151; and quality of life, 188; 189; scarcity, 218; services, 204, 207; as a social problem, 189; and socialism versus capitalism, 110; vernacular, 158; 166
energy indexes: in Tryon's work, 72, 75–76, 79; in Saha's work, 85, 93, 96; in Russia, 121
Energy's History, 3–4, 6, 11
Engels, Friedrich, 105, 108, 113, 115, 252; electrification 115, 122, 252; on industrial production, 113
entropology, 187

entropy, 6, 228; and feedback, 189; and the United States, 171
Ernesto Stumpf, Urbano, 25
Eskom, South African Electricity Supply Commission, 151
ethanol, 12, 21–22; anhydrous and hydrous, 23; cars and, 25, 29; vinasse and, 27
ethics of energy, 170
exhaustion. *See* fatigue
extinction, 210
extraterrestrial life, 5
Exxon, 191

fast-breeder nuclear reactors, 189, 198, 207
fatigue, 171, 177–78
Federal Power Commission (US), 152
female labor in Sudan, 164–65
Fiat, 25, 33
fifty-fifty profit-sharing model, 128–29, 142
Filho, Maurilio Biagi, 12, 22; Sr., 24; Filho, Pedro, 24; and Usina Santa Elisa, 24
firearms, 164
firewood, 82
First World War, 17, 23, 76; and energy conservation, 83; and food supply in Soudan, 158; and geologists in China, 68; the Great War, 54; Russia and Germany, 104–105
Fleck, Ludwig, 187
Fletcher, Horace, 180
flex cars, 30
Folha de São Paulo, 28, 33
food: preparation, 164; production, 215; shortages, 161. *See also* diet
forces and relations of production. *See* productivity
Ford, Henry, 22, 26
fossil fuels, 2, 7–12, 13–15, 18; boom, 211; fossil developmentalism, 14; fossil-

fueled domination, 13; fossil-fueled resistance, 13; mineral or fossil energy system, 7, 205; nonconventional, 207, 209; planetary consequences of fossil fuel use, 6, 224. *See also* coal, oil, ethanol
France, 22, 56, 158; Alsace-Lorraine, 66; French colonialism, 158–64
Franzese, Sergio, 170
Fressoz, Jean-Baptiste, 11
Friedman, Milton, 208
fuel mix, 23
fuel question (in Japan), 35, 36, 39–40
Fuel Society of Japan, 36, 37
Fukushima Daiichi nuclear disaster, 43
Fushun Prefecture, 12

Gallegos, Rómulo, 128
Gandhi, Mahatma, 90
Garavini, Giuliano, 14–15, 226
Geddes, Patrick, 96
Geisel, Ernesto, 21
gender, 17, 18; diversity, 190; and energy work in Soudan, 165; gendering energy technologies, 190
General Motors, 26
Germany, 2, 22, 42, 48, 51, 104, 110, 151; land of ersatz, 205; Ruhr Valley, 61, 66, 120; German geology, 58–59, 65; Allgemeine Elektrizitätsgesellschaft (AEG), 110, 151; German institutionalism (economic theory), 211; Nazi Germany, 208
Ghana, 18
Gilman, Charlotte Perkins, 175; and domestic energy, 175
Global South, 14, 124
Goeller, H. E., 217
Goethe, Johann Wolfgang von, 103
good Anthropocene, 212
Great Depression, 23, 78
green gas, 223
gross domestic product (GDP), 71

gross National Product (GNP), 34, 71, 188
growth models, 196
Guanying, Zheng, 59
Gulf Oil, 125
Gusterson, Hugh, 187

Hammond, Philip, 207
Han Dynasty, 65
Handler, Philip, 221
Hardin, Garrett, 202
Harrison Moore, Abigail, 18
Harte, John, 208
Hartwell, Robert, 9; on China's early industrialism, 57
Harvard University, 170
Harvard, Radcliffe College, 185
Hayek, Friedrich, 206
Heartland Institute, 211
heavy machine industry, Russia, 112
Herbert, Eugenia, 164
Holdren, Paul, 208
Homem de Melo, Fernando, 28
Hotelling, Harold, 73
household: electrification, 91; energy use, 39, 162–63; probity, 190
Hubbert's second peak, 211
Hughes, Thomas, 151
human capital, 203
human energy, 157
human-machine cooperation, 115
hunting, 164
hydraulic fracturing (fracking), 211
hydroelectricity 24, 30, 46, 85, 87, 92, 95, 110, 152; conservation and, 52; Dniepr Dam, 110; hydel power in India (hydroelectricity), 100; river management, 87–88; US Water Power Act, 152

Ibadan, University of, 214
ideational energies, 226
imagination, 220
Imperial Japanese Navy, 36

India, 10, 14, 84–102; British East India Company, 8; Indian National Planning Committee, 99; Indian National Science Academy, 87; Indian Science News Association, 87; Modi, Narendra, 2; *prana*, 16
infinitude, 219
institutionalist economics, 205; German institutionalism, 211
inter-industrial connection, 120
International Climate Change Agreements: Kyoto Protocol, 1; Paris Climate Accords, 1, 30, 55, 63; Intergovernmental Panel on Climate Change (IPCC): Sixth Assessment Report, 2, 63; IPCC and decoupling, 13
Inuit, 184
Iran, revolution, 24; Mossadegh's overthrow, 133
Israel, 18, 23, 213

Jamaica, 10
James, William, 16–17, 169, as an energy theorist, 169; critique of Henry Adams, 172–73; moral equivalent of war, 174; war against nature, 174
Japan, 12, 35–52, 224; and Britain, 39–40; coal import to, 37; coal question in, 46; fuel research in, 37–42; as "have-not-country," 35; Honshu, Kyushu, Shikoku, Hokkaido, 35; Japanese Bureau of Mines and the Mining Inspection Office, 53
Jevons, William Stanley, 42, 50
Johns Hopkins University, 125
Jordy, William, 6

Kaifeng, 9
Kardashev, Nikolai, 5
Kennosuke, Tsujimoto, 54
Kenny, Ramond, 74
King Hubbert, Marion, 211
Kluckhorn, Clyde, 186
Koopmans, Tjalling, 188

Korea, 35
Koselleck, Reinhart, 6
Kosoko, Oba, 146
Kōzō, Okunaka, 36
Krzhizhanovskii, Gleb, 14, 102, 103–11, 248
Kuznets, Simon, 205

La Peña, Carolyn de, 171
Lagos, Nigeria, 144; and the electricity questions, 156; Lagos treasury, 146; Lagosian elite, 148. *See also* Nigeria
Latour, Bruno, 187, 212
Lawrence Berkeley Laboratory, 194; conservation research at, 200
Lenin, Vladimir, 92–93, 101, 103–107, 109–10
Levasseur, Émile, 77. *See also* energy slaves
Lévi-Strauss, Claude, 186–87
lifestyle, 188; low-energy, 190
Limits to Growth, The, 202, 206, 212, 219, 271
Lomborg, Bjørn, 211
London, 9
Lotka, Alfred, 189
Lovins, Amory, 198
luxury, electric light as, 155
Lvov, Ukraine (Poland), 187

MacAskill, William, 212
machismo, 197–98
Madhusudan Dutt, Michael, 86
Mali (Soudan), 16
Malinowski, Bronisław, 187, 191
Malm, Andreas, 8
malnutrition, 210
Malthus, Thomas, 202; Malthusianism, 93
Mande people: cultural region, 157; farming rites, 158; social and environmental memory, 161–62
Manji, Yoshimura, 12, 35, 37, 39
manpower, 89
Maoism, 12

Marshall, Alfred, 44
Marx, Karl, 108–109, 113, 117–18; on steam power, 108–109, 122, 252
Marxism, 106, 108, 110, 213
master resource, 204
materialism, 171; matter, science of, 121
Meadows, Donella, 202
Medina Angarita, Isaías, 127
Mediterranean, 13
mental energy, 16, 172, 176
metal prices, 209
metaphysical healing, 180
methane (natural gas), 93
Mexico, 18
Middle East, militarization of, 191
Miller Beard, George, 171
Millet, 16, 158; as fuel, 165; *toh*, 162; *toh fanga*, 162
Miner's friend, 8
misdirected energy, 174
Missemer, Antoine, 13, 225
MITRE Corporation, 190
Mobil Petroleum Company, 66
modellers, 212
Moffat Yard & Company (publishers), 169
Molotov, Viacheslav, 107
Monteil, Charles, 163
Monterey, California, 193
moral energy, 170–71, 173
Morse, Chandler, 205, 217
Mumford, Lewis, 96
Murex shipping tanker, 10
muscle power, 95, 172
Musk, Elon, 213

Nader, Laura 17, 183–92, 227
Narodniki, 106
national energy economy, Soviet, 121
National Socialist Germany, 208
nationalization, 21
Navajo Nation, 186
Nef, John, 4, 9
Nehru, Jawaharlal, 99

Neo-Confucianism, 60
neo-Malthusians, 202–203
neoclassical economists, 212
neoliberalism, 205; neoliberal epistemology, 208, 211
neotechnic/paleotechnic eras, 96
Neto, Delfim, 28, 33
neurasthenia, 171
New Deal, 92; New Deal Prometheans, 207
New Scientist (magazine), 202
Newsweek (magazine) 206
Nigeria, 15, 144–54; power sector reform in, 152–53; and the World Bank/IMF, 152
Niigata, Japan (oil), 36
Nobutarō, Ishiwata, 48
nonconventional fossil fuels. *See* fossil fuels
Norberg, Johan, 211
Nordhaus, Ted, 212
North American Space Agency (NASA), 193
nuclear energy, 24, 30, 43, 94, 104, 111, 123, 188–91, 193–95, 197–98, 207, 225; age of, 123; in India, 94; in Japan, 43; safety, 194; in Soviet Union 104, 111, 225
Nunavut territory, 184
nutrition and energy, 178–79
nyama, 16, 226; *nyamaw*, 159, 164

Organization of the Arab Petroleum Exporting Countries (OAPEC), 205
Oaxaca, Mexico, 185
Odum, Howard, 189
oil: conservation and nationalization, 130; dependence and, 131; development and, 137; the "Devil's Excrement," 132; globally 131; governance, 124; Middle Eastern, 14; oil revolution of 1973, 125; oil weapon, 226; synthetic, 205; Venezuelan oil sector 125; and war, 104–105

OPEC, 14, 15, 23–24, 218, 226; and conservationist ideology, 132–33; creation of, 126, 134; and the Petroleum Pentagon, 136
Opium War, 8
optimization, 119; of societal energy, 173
organic energy, 86
organic refineries, 163
organicism, energy-economic, 107, 114
Osterhammel, Jürgen, 11
Ostwald, Wilhelm, 189
Otto engine, 25
Ottoman Empire, 10, 13

Paddock, Paul, 202, 216
Paddock, William, 202, 216
Palestine, 18
Paley Commission, 217
palm oil, 146
Papini, Florence, 181
Parthasarathi, Prasannan, 9
particularism, 184
Pastaza River, 187
peak oil, 210–11, 212
peat, 104, 225
Pena, Camilo, 33
People's Republic of China (PRC), 59
Pérez Alfonzo, Juan Pablo, 15, 125, 131
Persia, 10; Persian Gulf, 125, 218; oil from, 191
petro-state, 104–105, 126
Petrobras, 12, 34
Petróleos Mexicanos (PEMEX), 130–31
petroleum. *See* oil
Petroleum Pentagon, The, 126
Physics Today (magazine), 185
physics, 86–87; river physics, 92
pig iron, 51
Pinchot, Gifford, 52, 72
Pinkerton, Richard, 189
pipeline boom, 105
planetary system, 227
planning, energy economy. *See* central planning

playgrounds, 175
Plekhanov, Georgi, 106
Pomeranz, Kenneth, 9
Pooley, Gale, 211
Powell, John Wesley, 52
power. *See* energy
power fantasy, 16, 17
pragmatist philosophy, 173, 181–82
private-sector electrification, critique of, 150
privatization, of electricity, 14
Proálcool, 21, 26. *See also* ethanol; Brazil
productivity, 9, 72, 107; of labor, 98, 217; productive forces, 108–109; under socialism, 118–19
progress and energy, 225
progressive-era concerns, 129; 1901–1909 conservation movement, 72; reforms, 174
Project Camelot, 186
Promethean Marxists, 213
prorationing, 129; international prorationing, 133

Qatar, 223
qi (気) 16; *qihua* (氣化), 16
Qing empire, 56; dynasty, 65
quantification, 198

Rappaport, Roy, 183
rationalization, Soviet, 121
Reagan, Ronald 152, 191; governor of California, 186
Red Sea, 13
religion, 180; energy as the "American Religion," 191; Hayek and the market, 206; neoliberalism and faith, 213
renewable energy, 21–34, 55, 63–64, 189, 205, 223; solar power, 104, 189, 195, 199, 207–8, 220; wind power, 30, 51, 63, 97, 248. *See also* ethanol, hydroelectricity
RenovaBio (Brasil), 30
resourceship, 17, 203–204

revisionism, 18
revisionist economic history, 205
Rheinisch-Westfälische Elektrizitätswerke (RWE), 120
Richard Nixon, vice president, 129
Richthofen, Ferdinand von, 58; works on China, 58, 65; Richthofen's *China*, 67
Ridley, Matthew, 211
ritual energy, 159
rituals of reconciliation, 200
rivers: Damodar, India, 92, 101; Dniepr, Ukraine, 129; Pastaza, Ecuador, 187
Rockefeller, John, 124
Roosevelt, Theodore, 52, 171
runaway-bride crisis, 165
rural labor, 28–29
Russ, Daniela, 14, 225
Russia, 18; energy economy today, 104; fossil-fuel rent-based economy, 105; invasion of Ukraine, 64; Russo-Japanese War (1904–1905), 36

Sabino de Oliveira, Eduardo, 23
Saha, Meghnad, 14, 86
Sahel famine, 206, 214
Salk Institute, San Diego, 187
San Francisco Earthquake of 1906, 170
Sandwell, Ruth, 18
Sant'Ana, Carlos, 34
São Paulo, 24. *See also* Brazil
Schellenberger, Michael, 212
Schmitt, Carl, 93
Science (journal), 203, 204
Science and Culture (journal), 87
scientific history, 5
scientism, 184
Scott, Howard, 122
Second Industrial Revolution, 11
second wind, 171, 177
Second World War II, 138
Second World, 224
Senegal Valley, 163
Seow, Victor, 12, 224
sexism, 185

Shaku, 49
Shanghai Chamber of Commerce, 59
Shaw, Bernard, 177
Shia, Lebanon, 184
Shikai, Yuan, 61
shixue, 60
Sieferle, Rolf Peter, 7
Sierra Leone, 147, 150, 154, 157
Siguang, Li, 59
Simmons, Matthew, 211
Simon, Julian, 17, 203
Sivasundaram, Sujit, 9
slavery, 10, 85; energy slaves, 88–89; 98; and industrialism, 9–10
socialism: age of, 123; and great works, 117; socialized energetics, 115
Society of Producers of Sugar and Alcohol (Sopral), 33
socio-evolutionism, 192
Soddy, Fredrick, 189
soft energy paths, 198, 199
solar power potential. *See* renewables
Solger, Friedrich, 67
Solow, Robert, 205
Song Dynasty China, 9
Soviet Union, 14, 103–11; Commission for the Electrification of Russia (GOELRO), 103; Institute of Energetics, Soviet Academy of Science, 106; New Economic Policy (NEP), 112; Soviet Planning Agency (Gosplan), 103
Steinmetz, Charles, 121
space colonization, 204
Spaceship Earth, 213
St. Petersburg Technological Institute, 105
Stalin, Joseph, 15, 104, 106–107, 110
Standard Oil, 134
Shell, Royal Dutch, 125
Stanford University, 170
starvation, 210
Statoil, 191
steam power, 8–10, 224; age of, 123; steam power in Lagos, 145–46

Stinnes, Hugo, 120
strikes, 44, 127, 239; and coal, 44; *grevinhas* in Brazil, 28–29, 239
Structural Adjustment programs, 152
studying up, 186
subaltern perspective, 90
subjectivity, 190
Suez Canal, 10, 13
sugar: Brazilian sugar crop acreages, 26; cane, 26; industry, 24; "an ocean of sugarcane," 27; pollution, 27
Sun, 219–20, 227
supernatural abilities, 160
Sweden, power in, 97
Switzerland, 188; power use in, 97
Syria, 23

taboos, 193
Taishō (nonexistent dates), 48
Tang Zonghai, 16
Tanzania, 18
Tariki, Abdullah, 129, 133–134
Tata family, 91
Taylor, Fredrick, 115; Taylorism, 171
technocrats, 12, 73, 93, 122, 127, 129, 133, 136
Tennessee Valley Authority (TVA), 92, 101
Texas Railroad Commission (TRC), 129, 136
Thammasat University, 186
Thatcher, Margaret, 152
Thermodynamics, 6, 90. *See also* entropy
Third-wave energy anthropology, 184
Three Rapid People, 16, 157
Tierney, John, 211
Tokyo Imperial University, 39
Tonight Show, The, 202
totem, 190
Touré, Samori, 164
Trans-Atlantic slave trade, 146
transition, 84; critique of, 224; wood fuel, cooking, and steam power, 158
transportation, 83

Travélé, Moussa, 16, 157
Trobriand Islanders, 187, 191
Tryon, Fredrick, 13, 72
Tupy, Marian, 211
Turnbull, Thomas, 17, 227
Twagira, Laura Ann, 16, 226

United African Company (Unilever), 150
United Arab Republic, 133
United States of America, 11, 23, 30, 39–40, 42, 50, 52–53, 56, 76–77, 80, 90, 92, 122, 124–25, 127, 129, 134, 136, 139, 141, 147, 152, 170, 210, 216–17, 223, Bureau of Mines, 49, 53; coal economy in, 47–48; Federal Water Power Act of 1920, 152; Geological Survey 53; Inland Waterways Commission, 52; Interstate Compact to Conserve Oil and Gas, 130; power use in, 97; Progressive Era, 13, 72
Urbana-Champaign, University of Illinois, 203
usinas (mills), 32
utility company planners, 198

Van Hise, Charles, 52
Venezuela, 15; coup of 1946, 128; fiscal sovereignty, 127; and Iran, 133; *La Apertura* (The Opening), 125; strike of 1936, 127; Venezuelan Oil Company (CVP), 130–31, 140
vernacular thermodynamics, 16
Vettese, Troy, 17, 227
Victoria Falls and Transvaal Power Company Limited, 151
vinasse, 27
Vincente Gómez, Juan, 125, 127
Volkswagen, 25, 26, 33

Waldheim, Kurt, 214
Ware, Helen, 214
Watt, James, 8, 10, 75
Weinberg, Alvin, 207, 217
Wells College, New York, 185

Wenjiang, Ding 13, 57; *alias* V. K. Ting, 70
Wenlie, Tian, 58
West India, 10
White, Leslie, 5, 17, 183
Williams, Eric, 9
Wolof folk tales, 163
Wong, W. H., 68
wood fuel and cooking, 158
Woolgar, Steve, 187
work, 4, 6, 89
World Energy Council, 73
Wright, Rebecca, 16, 226
Wrigley, Anthony, 7
Wu, Shellen, 13, 225

Xi Jinping, 63

Yangzi Delta, 9
Yergin, Daniel, 9
Yiou, Zhang, 56
yoga, 173
Yom Kippur War, 23
Yoruba, 147

Zahlé, Lebanon, 185
Zaporishia, Ukraine, 110
Zapotec, Mexico, 184
Zimmermann, Erich, 205

The authorized representative in the EU for product safety and compliance is:
Mare Nostrum Group B.V.
Mauritskade 21D
1091 GC Amsterdam
The Netherlands
Email address: gpsr@mare-nostrum.co.uk

KVK chamber of commerce number: 96249943